FINDING TROUT
IN ALL CONDITIONS

**A GUIDE TO UNDERSTANDING NATURE'S FORCES
FOR BETTER PRODUCTION ON THE WATER**

BOOTS ALLEN

WESTWINDS
PRESS®

PRUETT
THE PRUETT SERIES

Library of Congress Cataloging-in-Publication Data

Names: Allen, Joseph Boots.
Title: Finding trout in all conditions : a guide to understanding nature's forces for better success on the water / Boots Allen.
Description: Portland, Oregon : WestWind Press, [2016] | Includes bibliographical references and index.
Identifiers: LCCN 2015041025 | ISBN 9780871083289 (pbk.) 9781943328413 (e-book) 9781943328758 (hardbound)
Subjects: LCSH: Trout fishing. | Fly fishing.
Classification: LCC SH687 .B64 2016 | DDC 799.17/57—dc23 LC record available at http://lccn.loc.gov/2015041025

Edited by Todd Berger
Designed by Vicki Knapton

Published by WestWinds Press®
An imprint of

GRAPHIC ARTS
BOOKS®

P.O. Box 56118
Portland, Oregon 97238-6118
503-254-5591
www.graphicartsbooks.com

This work is dedicated to Faith and Mason,
the two most important people in my life.

CONTENTS

ACKNOWLEDGMENTS

This book comes out of several decades on the water. Much of this time has been spent with family, friends, mentors, and heroes of mine, all of whom have shared their knowledge and observations. Their support over the years has meant so much to me. There are many who deserve special mention.

My grandfather, father, and uncle were all longtime guides and outfitters in the Greater Yellowstone area. They introduced me to fly fishing at a very young age (perhaps too young), and they taught me to pay attention to everything.

My brother Richard and cousin Tressa have been fishing companions of mine since I was in elementary school. Both guided for a number of years and could pick up on things happening on the water that I was oblivious to. We still get to fish together every year, and I continue to learn from them.

Fly-fishing writers Paul Bruun, Scott Sanchez, and Bruce Staples have shared their passion for the craft, as well as a ton of tidbits that have made the process of writing a book a little easier. All three are heroes of mine. Those times I have had over the years to sit down and talk with them have been among the most cherished moments of my life.

A number of smart and influential individuals in the sport of fly fishing have shared their experiences with me. Much of what I include in this book is inspired by my discussions with them. These individuals include Carter Andrews, Tim Brune, Jean-Williams Bruun, Ken Burkholder, Darren Calhoun, Jeff Currier, Mike Dawes, Mike Dawkins, Will Dornan, Kelly Galloup, and Craig Mathews.

Being on the water guiding or fishing close to two hundred days a year gives me ample opportunity to interact with fellow guides. They are the true sentinels of the water and the ones whom I learn the most from year in and year out. Those from whom I have gained the most knowledge include Kevin Brazell, Matt Breuer, Dean Burton, Josh Cohn, Ed Emery, Kurt Hamby, Jeff Hanson, Josh Heileson, Max Mamaev, Shannon McCormick, Steve Mock, Jaason Pruitt, Jim Reetz, Keith Smith, Scott Smith, Patrick Straub, Sue Talbot, Anya Tobie, Oliver White, Benny Wilson, and Trevor Wine.

Lastly, I would like to thank the editing team at Graphic Arts Books for their patience, enthusiasm, and thorough attention to detail with this project. Doug Pfeiffer, Kathy Howard, and Vicki Knapton are amongst the best in the business. Thanks for all your hard work.

INTRODUCTION

Fly fishing is a game full of ancillary components. Studying them allows us to become better at catching fish as well as learn more about the natural world. Entomology is a prime example. Studying insects, be they terrestrials or waterborne, can assist anglers in determining what patterns to use during a certain part of the day or season, what water types to target, and how best to fish and present the fly. But entomology

Understanding the vast components of fly fishing, from holding water and trout behavior to entomology and external factors, makes all of us better anglers.

also gives fly fishers the chance to learn much more about the ecology of streams and lakes. This kind of synergy exists in fly fishing just as much as it exists in other outdoor sports like skiing, surfing, or kayaking. In fact, this is synergy at its best—we become better anglers by learning more about the waters we fish, and we learn more about the waters we fish by becoming more knowledgeable anglers.

The amount of knowledge a fly fisher can absorb is easily illustrated through a review of videos, fly-fishing magazines, and angling websites. Dig deep into these resources, and it is amazing what you will find. There will be topics such as the biology and behavioral traits of trout. You will also find articles and videos on hydrology and its influence on holding water for trout. You can also find gear- and tackle-oriented themes like rod and line technology, leader construction, and the design of watercraft used to stalk trout. And, of course, there is the ever-present subject of casting and the biomechanics behind this all-important piece of fly fishing. The possibilities for greater knowledge are truly immense.

Now consider the diversified nature of the sport as it exists today. Trout and the cool freshwater streams and lakes in which they live are at the heart of fly fishing and always will be. Yet fly fishers now have the ever-increasing popularity of saltwater angling, fishing for warm-water species like bass and carp, and swinging flies to anadromous fish like steelhead, salmon, and sea-run brown trout. Stalking these game fish allows anglers to learn more about the sport and the various ecosystems in which we play. Those who primarily target trout can take particular aspects of saltwater, warm-water, and anadromous fishing and apply them to the freshwater streams and lakes they fish. Today, we not only use fly patterns from saltwater and sea-run angling to fish for trout, we also apply some of the tactics from these types of fishing when we are going after trout.

Having intimate knowledge of these aspects of fly fishing not only makes us better fly fishers, it also make us smarter fishers. We have the ability to go into our favorite waters, or even waters we have never before fished, better equipped to deal with the task of catching more and bigger trout. Our confidence can reach off-the-chart levels.

Many strategies, tactics, and patterns used in saltwater and anadromous fly fishing can be applied to trout. Having experience with the diverse aspects of the sport can take your angling to a whole new level.

Yet no matter how prepared we think we are, Mother Nature tends to have a way of throwing curveballs at us. We are left scratching our heads at a sudden part of the day, or an entire day, or an entire week, of unproductive fishing despite our thorough knowledge of the cast, the presentation, the holding water, the available forage, and the patterns we are using. These curveballs can come in many forms, anything from weather and water temperatures to pH levels and moon phase. These facets of the sport are apt to be the last ones we consider when fishing.

Some fly fishers believe that low atmospheric pressure and precipitation work together to produce the emergence of numerous mayfly species. Knowledge of these climatological factors can help the angler better understand the waters they fish.

Gaining greater knowledge about these factors, sometimes referred to as *external* factors, *natural* factors, or *ecological* factors, can seem like a lot of work. We already feel obliged to know everything about aquatic insects or where fish are located on the water we are fishing. Isn't delving into the mysteries of weather, climate, and water overkill? Does it really matter in the grand scheme of your day on the water? In my many years of fishing and guiding, I would answer these questions with a definite *maybe*. I know many fly fishers who don't have even the basics of entomology down. Nor do they know everything there is to know about the rod they use. Nonetheless, these anglers do just fine no matter where they fish.

I find that the most skilled fly fishers have a sincere desire to learn as much as they can about the sport. They want to learn more not only because it makes them better anglers, but because the process of learning more about fly fishing is satisfying in and of itself.

In this book, I demonstrate that external factors matter and should be considered an important component of fly fishing. Many fly fishers ignore external factors, and much of this is because these elements have been ignored by those of us who write about the sport. At best, they are glossed over. Two of the most recent exceptions are Taylor Streit's *Instinctive Fly Fishing* (Lyons Press) and Denny Rickard's *Fly Fishing Stillwater for Trophy Trout* (Stillwater Productions). This book expands upon this overlooked aspect of the sport.

I begin *Finding Trout in All Conditions* by examining climatological and weather-related factors such as barometric pressure, precipitation, air and water temperature, wind, sunlight, and cloud cover. These elements are often related, with one impacting the others and vice versa. I investigate each separately, focusing on how particular changes—the rise and fall of a barometer or the thermometer for instance—influence the behavior of trout and the various types of food they consume.

After examining these aspects, I turn my attention to water-related factors, including water levels on lakes and streams, water temperatures, pH level, dissolved oxygen content, and specific conductance of the water being fished. Outside of water levels, these factors are often neglected by the fly fisher. I illustrate just how important these factors can be and how understanding their significance can be a big advantage to the angler.

After investigating water-related factors, I turn to one of the most mysterious influencers in the world of fishing: moon phase. Lunar influences on fish are something most anglers recognize but do not completely understand. Some say a full moon renders productive trout fishing impossible. Others say that fishing during a full moon is full of potential, but success comes from focusing attention on specific periods of the day during the full moon, and these periods can vary from one river or lake to another. I explain my experience and the experience of other expert anglers with moon phase when fishing for trout. As one renowned stillwater angler has said, "Trout can and are caught during a full moon, but only when it is right for them, not me."

In 1998, as I started my eighth season of guiding, I began keeping a journal to record daily events and the productivity of each outing. About six years later, I began to use a standardized fishing journal where I could record a number of variables and the results of the day. These journal entries now total well over 1,600 cases (days on the water). I enter this data into a statistical software package that allows me to analyze the information. It is this data, coming from streams and lakes all over North

Since 1998, I have used journals to keep a record of each outing. Standardized logging systems like Rite in the Rain allows users to better record and analyze their data. From this data, the fly fisher can detect trends in external factors that impact trout.

America, as well as my discussions with a number of veteran anglers, that informs this book.

Throughout each chapter, I use the terms *external*, *ecological*, and *natural* interchangeably when explaining these factors. The reason for this is that these terms are used with equal abundance in the literature (if they are mentioned at all) to define issues relating to weather, climate, water, and moon phase. Fly fishers who examine these factors have yet to decide on one all-inclusive term. This book will not do much to move the literature in that direction. It will, however, do much to increase the angler's understanding of these factors and how they can impact fly fishing.

CHAPTER ONE

BAROMETRIC PRESSURE

There was no better place for an outdoor-loving kid like myself to grow up than in the mountains of the Yellowstone region. What made my early years even more fulfilling was the mentoring I received from a family of outfitters. Every year I could count on excursions to a variety of rivers and lakes, fishing for all kinds of trout. Autumn meant splitting my time between fishing for big browns and cutthroats, and hunting for ducks, grouse, pronghorn, and elk. And while winter was dominated by skiing, I always made time to accompany my father to our high-country lakes to ice fish for lake trout. It is hard to imagine a better upbringing for a youngster.

What stands out about these early years are my family's instincts regarding the natural world and the changes that occur within it from day to day, if not hour to hour. My grandfather's intuitive nature comes to mind in this regard. He had come to the valley of Jackson Hole, Wyoming, as a teenager in the 1920s. By the time he was in his early twenties, my grandfather was guiding on nearby lakes and rivers. When World War II ended, he purchased a few parcels of land in the town of Jackson, built a fly shop, and began outfitting full-time. It was a satisfying life, but a hard one that left his body sore and aching for the rest of his years.

One of my grandfather's main physical complaints was a painful knee caused by decades of scrambling over the cobblestone-strewn riverbeds during his forty-plus years guiding. When I joined him for trips to Jackson or Jenny Lakes in Grand Teton National Park, he could sense the

The L on weather maps indicates the position of a low pressure system. Such systems are associated with thunderstorms and inclement weather, as this map shows. (Image courtesy National Oceanic and Atmospheric Administration)

approach of storms hidden behind the towering mountain range to the west by the tightening of his knee joint.

"How do you know there is a storm on the way?" I asked him one day as he fired up his Evinrude motor to return to the boat ramp.

"That bad knee of mine always acts up when a storm-is-a-comin'," he replied. "Each storm means the pressure is dropping, and when that happens, my knee stiffens up like a corpse."

Stories like this illustrate the impact barometric pressure has on the world around us. It influences every living creature, including fish.

There are many climatological events that accompany changes in barometric pressure. It impacts (and is impacted by) wind, precipitation, air temperatures, and cloud cover. In coming chapters, I will discuss these factors in detail, but it is important to start with that which influences them all: barometric pressure.

BAROMETRIC PRESSURE DEFINED

Barometric pressure (sometimes referred to as atmospheric pressure) is the force exerted on a surface by the weight of air above that surface. We don't often think of air or other gases as having weight, but the fact is that air, and the effect of gravity on it, does weigh down on earth, the creatures dwelling upon it, and the various types of surfaces and substances upon the planet, including water. The standard barometric pressure at sea level is 14.7 lbs/square inch, but barometric pressure can be impacted by geography and elevation. As one moves higher in elevation, the standard barometric pressure decreases because there is less air, and thus less pressure from weight of air, to weigh down on an object or surface. At 1,000 feet above sea level, the standard pressure is 14.1 lbs/square inch. Four miles above sea level is considered the halfway mark in our planet's atmosphere, where half of the earth's atmosphere is above this point and half is below. At the four-mile mark, the standard pressure is close to seven.

While altitude plays a fundamental role in barometric pressure, fly fishers, and most individuals who play in the outdoors for that matter, are more concerned with how barometric pressure influences the weather and how both of these factors taken together influence trout and the waters they call home.

When meteorologists consider barometric pressure, they think of it existing in two states: one of high pressure and one of low pressure. But

A clear, sunny sky typically accompanies high-pressure systems. Fishing can be good, but these conditions can also present challenges.

they also recognize transitional states, such as moving from high pressure to low pressure or from low pressure to high pressure. From a weather standpoint, barometric pressure can be influenced by one or more of the following:

1. **RISING AIR:** Winds that move horizontally do so at speeds far greater than winds that move vertically. When vertical wind speeds increase dramatically, it leads to decrease in downward pressure upon the earth's surface. The result is a decrease in barometric pressure. A good example of this is the upwelling of wind in intense thunderstorms or tornados.

2. **WARMING AND COOLING AIR:** Warm air is less dense than cold air. As such, warming air rises and relieves atmospheric pressure on the earth's surface. Because cold air is dense, its weight will put more pressure on the earth's surface. A good example of this is temperature inversions that occur in mountainous regions, where cold air at a low elevation can be trapped by warm air at higher elevations. In my hometown of Jackson Hole, Wyoming, it is possible to have temperatures around 20 degrees at the 6,200-foot level while there are temperatures closing in on 40 degrees at the 10,000-foot level during the winter months.

3. **MOIST AND DRY AIR:** Moist air is less dense than dry air. As with cold air and warm air, the density of dry air puts downward pressure on the earth's surface, while moist air lessens this downward pressure. An example of this is moisture-laden clouds that appear high in the atmosphere. Only when these clouds build up significant mass will they put downward pressure on the earth's surface in the form of thunderstorms.

4. **APPROACH OF LOW- OR HIGH-PRESSURE TROUGHS:** These can be thought of as cold fronts or warm fronts. A low-pressure trough can cause air to rise from the surface to a higher part of the atmosphere, air density to decrease, or air mass to decrease.

All of these cause a decrease in the pressure on the surface of the earth. Conversely, a high-pressure trough can cause air to drop to the surface from a higher part of the atmosphere, air density to increase, or air mass to increase. This can cause an increase in the pressure the atmosphere imparts on the earth's surface.

5. **THE DEEPENING OR LIFTING OF A TROUGH OR FRONT:** The deepening of a low-pressure trough signifies the intensification of rising air, decreasing air density, or decreasing air mass. The lifting of a high-pressure trough signifies the intensification of air dropping from the atmosphere to the earth's surface, increasing air density, or increasing air mass. In more simplistic terms, a deepening of a trough signifies the amplification of a cold front and the lifting of a tough signifies the amplification of a warm front.

Weather accompanying high barometric pressure is often sunny and warm, as well as windless. Low barometric pressure can be accompanied by cool, cloudy weather with precipitation. A transition from low pressure to high pressure is usually accompanied by clearing, drying skies and warming air temperatures. A transition from high pressure to low pressure can be accompanied by increasing clouds and a drop in air temperature with a chance of precipitation. But always remember that these are general trends in weather with various types of atmospheric pressure. Anomalies, as well as unique geological features, exist that, at times, may lead to high pressure when all the climatological signs point to a lowering of barometric pressure and vice versa. This is what makes weather so fascinating.

Cold, wet weather typically defines low-pressure systems. This day on one of Idaho's desert lakes was a tough one for hookups.

Barometric pressure is a complex and convoluted piece of science, and getting lost in the details can be pretty easy.

BAROMETRIC PRESSURE AND FLY FISHING

Our atmosphere exerts pressure on water just as it does on land. All creatures living in water feel the effect of that pressure. This includes fish.
Like so many types of natural phenomena, we don't fully understand
how barometric pressure impacts fish behavior. With water being eight
hundred times denser than air, changes in the pressure of our atmosphere could produce an additive impact on water-dwelling creatures
that are known to be sensitive to pressure. Fish can sense changes in the
barometric pressure in their air bladder. Some saltwater game fish like
dorado, mackerel, and wahoo have small bladders in comparison to their
body sizes. As a result, many biologists believe that they are less affected
by changes in barometric pressure. Other game fish have bladders that
are considered large for their body size. Trout fall into this category.
These fish are more impacted by barometric pressure.

When barometric pressure drops, the air bladder in trout will
expand because of the lack of pressure on their bodies. Freshwater biologists believe that this expanding air bladder places uncomfortable pres-

*Some game fish have large air bladders that are impacted greatly by changes
in barometric pressure. Saltwater fish like bonefish fall into this category. So,
too, do trout.*

sure on the organs of trout. Trout can relieve this pressure by descending lower in the water column. The pressure of deeper water replaces the pressure of the atmosphere. As barometric pressure transitions from low to high, the air bladder will contract, allowing trout to rise back to a higher part of the water column.

All of this would suggest that the transition from high barometric pressure to low barometric pressure would produce slow fishing because trout are experiencing discomfort from their expanding bladders and that during low-pressure periods anglers must fish well below the surface because they are holding at a deeper piece of the water column. This would seem logical, and no doubt most of us have experienced this kind of fishing under these conditions. But many of us have also experienced blockbuster fishing when there is a low-pressure system moving in and when a low-pressure system has stabilized in a specific location.

We know barometric pressure impacts fish and, thus, our ability to rack up strikes and hookups. The truth is, however, that the quality of fishing can vary widely during each type of pressure phase. What is more, we get locked into thinking that each type of pressure will guarantee specific results, be they fast or slow. And what one fly fisher expects can be completely different compared to

The impact of barometric pressure on trout can vary from lake to lake or stream to stream. How lake trout in the Yellowstone Region react to pressure changes can be completely different from how it affects the same species on lakes in the Sierra Nevada.

what another fly fisher expects. When I was performing research for this book, I came across Dr. David Ross's *The Fisherman's Ocean* (Stackpole). In his book, the marine scientist states that barometric pressure does not impact saltwater fish in ways that should be considered crucial by the fly fisher. At the same time, Spud Woodard of the Georgia Department of Natural Resources Coastal Resources Division claims that fish are most comfortable during a stable high-pressure system and will change their feeding patterns noticeably during and after a transition from high pressure to low pressure.

Opinions and theories abound when anglers discuss how barometric pressure impacts trout fishing. Discussions become even more convoluted when fly fishers begin to bring up specific waters. I have heard from a longtime Henry's Lake angler how the barometer influences his outings, and it is completely different from the experiences of another fly fisher I know who frequents Eagle Lake in northern California.

These are just theories and opinions. Nonetheless, I am in the company of many other fly fishers who have discovered trends between types of barometric pressure and the fishing we experience:

High Pressure

As discussed earlier in this chapter, high pressure is characterized by clear skies with significant sunlight. This does not mean that temperatures are necessarily warm. Some high-pressure periods in the winter months can be extremely cold. The general reading of a barometric pressure gauge during periods of high pressure is at or above 30.3 on most waters that I fish. Keep in mind that on high-elevation lakes and streams (those at 5,000 feet or higher) a gauge can read slightly lower—anywhere from .02 to .10 in my experience. Fishing during periods of high pressure can be fairly consistent. However, this does not mean that the fishing is easy. On streams, I do not find a lot of action in shallow water types such as shallow riffles, flats, and seams. Instead, they are located in deeper pieces of the water column or in *tight* to structure and banks. For those who fish storied dry fly streams

High barometric pressure does not always mean warm air temperatures. Winter fishing under clear skies can also be accompanied by cool temps. And, yes, the fishing can be quite good.

During periods of high pressure, I find that trout hold very tight to their holding water. When fishing riffles, seams, banks, or structure, the best action will be at the head of the feature forming the current break. Fish the entire piece of water, but concentrate on the upstream portion.

and lakes, high-pressure periods can offer good fishing, but it is not typically blockbuster.

When fishing deeper parts of holding water during high-pressure periods, using nymph rigs and streamers (heavily weighted or fished with a Type III or heavier sink tip) is a no-brainer. This does not mean that dry flies are not in the cards. There can be very good fishing during high pressure as long as the fly fisher is serious about the presentation. Razor-tight placement along banks and structure is necessary. In my experience, trout will not move more than a foot from their cover, depending on stream gradient and current velocity. Razor-tight placement is also required when fishing deep riffle pools formed by a steep shelf drop-off. The fly should be placed on the shelf and allowed to drift into the pool. Trout are usually holding directly at the head of the pool where the shelf begins its descent. This line of transition between the shelf and the pool is where most strikes will occur.

Stillwater can also offer consistent fishing during periods of high pressure. Just as on streams, it is best to target structure such as vegetation and deeper holding water, particularly near drop-offs. I do not experience consistent activity on flats unless they have depths approaching six feet or more. Fishing submerged bars can offer some of the best fishing on most lakes during high-pressure periods.

Transition: Rising Barometric Pressure

No barometric pressure period offers more opinions about fishing productivity than transitional phases. While I was guiding in Argentina several years ago, a very astute angler held the opinion that transition periods were the toughest conditions one can fish in. At the same time, one of my favorite guests whom I guide every year in Wyoming and Idaho is convinced that transitions were the best time to be on the water.

During a period of rising barometric pressure, I typically experience fishing that is somewhat similar to what I observe during high-pressure phases. Shallow holding water and the upper parts of the water column can have activity, but as the transition continues, activity here begins to slow and deeper lies, as well as banks and structure, begin to produce. I have noticed that fishing slower water with dry flies and nymphs produces better than fast currents during transitions to high pressure. Patterns that are fished with animation, such as streamers or soft-hackle nymphs, get more action when they are fished more slowly. Something like one-third to two-thirds speed is a good rule of thumb. Again, this might be due to the impact of high pressure on the swim bladder. Moving quickly to a fast-moving fly in a fast current just may be too uncomfortable, or perhaps too disorienting.

Animated patterns like Mike Mercer's Glasstail Caddis Pupa produce best for me when barometric pressure is transitioning from low to high. I use slightly slower retrieval speeds, and target slower runs and current margins.

The impact of rising barometric pressure on streams is mirrored on stillwater for the most part. I generally notice a transition in action from shallow flats, the upper parts of the water column, and littoral zones to deeper portions of flats, deeper parts of the water column, and drop-offs. The movement of my fly, however, is not slowed down at all like it is when I am fishing streams. I tend to move my fly, be it a nymph or a bait-fish imitation, with a speed that mimics that of the natural food I am imitating.

Moderate Pressure

Meteorologists generally consider the atmosphere to be in states of either high pressure or low pressure, with transitional periods as the atmosphere attempts to balance itself. Nonetheless, periods of moderate pressure do occur. On the streams and lakes of the Rocky Mountain West, I define moderate pressure as any reading on a barometer from 29.7 to 30.2. This can be slightly higher at lower elevations, although the differences can be so slight that it might be splitting hairs. Moderate pressure is usually characterized by stable weather that might include passing clouds or high cirrus or altostratus clouds. It could be described quite simply as fair weather.

Periods of moderate pressure are short-lived and tend to provide the most inconsistent fishing in my experience. This inconsistency is not only in terms of the patterns I use, it also goes for the presentation and the holding water trout are using. I have had blockbuster days on trout streams of all sizes and gradients with surface patterns and nymphs during moderate-pressure phases. But I have also had those days when it is a struggle, and I must employ a variety of tactics, such as fishing streamers with fast retrieves, swinging nymphs through the tails of riffles and seams, or skittering big surface attractors. Generally none of these tactics produce consistently. You just have to keep casting and pick up as much action as you can. If you stick with just one tactic, production might be far less than if you try everything you can.

The same can be said for stillwater fly fishing. I can remember a two-day, midseason outing on Montana's Hebgen Lake. Both days had almost identical weather conditions and moderate barometric pressure.

The atmosphere is normally in states of either high pressure or low pressure. Because of this, extended periods of moderate pressure do not often occur. When they do, fishing can be inconsistent at best, especially of stillwater.

On the first day, we were on the water a little after 10:00 A.M. and had excellent fishing on chironomid and *Callibaetis* larva imitations. We even had a fun period with surface feeders for about an hour after 4:00 P.M. The next day we were on the water about an hour earlier and fished until just before dusk. The fishing was decent, but my companions and I agreed that it did not come near to matching what we experienced the day before. In fact, most of our action came on nymphs that we stripped with erratic movement on flats and drop-offs.

Despite the inconsistency, periods of moderate pressure should not be avoided by anglers. Fishing can be as good as any other phase of barometric pressure. But if there is little cooperation by trout, my suggestion is to start experimenting with a variety of flies, tactics, and holding-water types. I have had only a few days where the fishing could be described as abysmal during moderate-pressure phases. That is because when it seemed slow, I pull out every card in the deck to increase my hookup rates and look for patterns of consistency. And remember: moderate pressure does not happen that often. And when it does, it doesn't last very long.

Transition: Falling Barometric Pressure

Some of the best days I have ever had on trout streams and lakes have been when the barometric pressure is falling. It seems that the faster it falls and lower it falls, the better the fishing is. It can be a feeding frenzy both on the surface and below. This is not just my experience. It is the experience of many longtime fly fishers who spend a significant amount of time going after trout. Browse the works of some of the best fly-fishing writers, and you will find they generally agree.

Falling barometric pressure is generally characterized by cooling temperatures, increasing cloud cover, and the potential for precipitation. There are a number of theories as to why fishing productivity can be so good when barometric pressure is falling. There is a belief that trout can feel the impending change in pressure from high to low and that, in preparation for the discomfort on their swim bladder, they will begin to

During periods of dropping atmospheric pressure, a number of patterns and tactics can be effective. I often turn to streamers. My belief is that the general intensification of insect emergences causes baitfish to feed more near the surface. Large trout typically respond by going into a feeding frenzy on both emerging invertebrates and any baitfish that get in their way.

Transitions from high to low atmospheric pressure can produce good action near the top of the water column on stillwater flats. Scott Sanchez's Fur Damsel is a lightly weighted damselfly larva imitation that works well in such conditions. Slow retrieves with variable movement is crucial.

feed aggressively. There is also a line of thought that on many streams, certain kinds of aquatic insects become active and move into an emergent phase when barometric pressure falls and that trout will have greater access to these insects as forage (I will discuss this in greater detail in Chapter 2). These reactions on the part of both the fish and the foods they eat probably work in a synergistic manner to produce excellent fishing.

What makes fishing so exciting during periods of falling barometric pressure is that wide varieties of holding water, as well as the entire water column, are in play. If I am fishing below the surface, I often turn to streamers because feeding is so active and trout tend to be sensitive to anything that moves. Trout will hit streamers as a food source or as an intruder who is violating its lie. Trout can often hold near the surface in riffles, seams, and eddies to feed on emerging aquatic insects. This makes falling pressure transitions ideal for fishing dry flies.

On lakes, fishing during a drop in barometric pressure is similar to what I experience on streams. Fishing deep in the water column on submerged bars or at the edge of drop-offs can be as good. But what makes lake fishing fun during a fall in pressure is the amount of activity on shallow flats and near the surface. There is nothing like stillwater fish hitting your fly at or near the surface. This activity can become so intense that trout feeding just below the surface produce very evident breaks on the surface. Retrieval speeds can play a role. I stick with retrieves that mimic the natural movement of the aquatic insects I am imitating when I am nymphing. With streamers, erratic or high-speed retrieves tend to out-perform others. Perhaps speed and erratic movement gets the attention of trout that have so much going on in their world during periods of falling barometric pressure. With damselfly- and dragonfly-larva imitations, I use a retrieve that incorporates a quick, six-to-eight-inch strip followed by four to six slow hand-twist or pinch retrieves. This style of retrieve is suggestive of the natural movement of these insects.

Low Pressure

If falling barometric pressure produces the best results when I am fishing on trout streams and lakes, periods of low pressure are a close second. Some well-known fly fishers do not share this sentiment. For me, how-ever, a low-pressure period is a great time to be on the water. It can be as inconsistent as what I experience during periods of moderate pressure, but when it is on, it is *really* on.

Low pressure is defined as a reading on a barometric pressure gauge below 29.7. It is characterized by cool or cold temperatures, cloudy weather, and precipitation in the form of rain, sleet, or snow. Fishing can be uncomfortable. Yet when the fish are feeding voraciously, it can be worth the discomfort.

Just as with falling barometric pressure, activity can take place in a variety of holding-water types and throughout the entire water column on trout streams. I tend to focus my attention below the surface with streamers and nymphs because during a day of consistent low pressure, the subsurface activity tends to be consistent as well. Surface fishing offers less-consistent action during low-pressure phases, but I experience at least sporadic activity with top-water patterns during most days of low pressure. When this activity does occur, it is typically during a certain one-to-three-hour period during the day. These trout are often feeding in a specific type of holding water. This could be at the tail of a seam or at the margin of the main current line flowing through a riffle pool.

Stillwater fishing during periods of low pressure can also be worthwhile. However, there are some noticeable differences with what I have observed on trout streams. Perhaps the most abrupt difference is the substantial lack of activity at or near the surface. The lower the pressure drops, the less activity there will be. This lack of activity extends down to about the six-foot level below the surface on most of the lakes I fish. Where I have most of my action is on deeper features like submerged bars or drop-offs. I can recall a number of frigid days with sleet and snow in mid-autumn on Lewis Lake in Yellowstone National Park when fishing was the best it had been all year. What made the difference was my use of a Deep 4 full-run sinking line and a Mohair Leech retrieved with a slow hand-twist retrieve. Eighty percent of my hookups occurred six feet or more below the surface.

This nice brown trout was picked up during a day of stable low pressure on Lewis Lake in Yellowstone National Park. Guide Joshua Cohn was slowly retrieving Callibaetis *nymphs at the eight-foot level in the water column, picking up several fish that day.*

Stillwater bodies like Henry's Lake in Idaho and the lakes of Yellowstone National Park offer good fishing at times during low barometric pressure, but it is important to note that these waters are considerably high in elevation (above 6,400 ft.) compared to many other trophy trout lakes. I have had the opportunity to fish lakes at much lower elevations, such as Oregon's Upper Klamath Lake (4,140 feet) and Washington's Crescent Lake (580 feet). When I have discussed the effect of low barometric pressure on these lakes with seasoned anglers who fish them regularly, they have explained to me that it is one of the poorest pressure phases one can fish. In fact, stillwater fly-fishing legend Denny Rickards say in his book *Fly Fishing Stillwater for Trophy Trout* that the lower the pressure drops below 29.7, the better it is to "stay home and tie flies." He fishes lower-elevation lakes on the Pacific Coast dozens of times a year. They are his home waters in every sense of the word. This leads me to believe that the lower the elevation a body of stillwater is located, the more negatively impacted it is by low barometric pressure. At the same time, this theory of mine is countered by other experienced anglers from the Rocky Mountain region who contend that it is high-elevation waters that are more negatively affected by low barometric pressure.

A FINAL WORD ABOUT BAROMETRIC PRESSURE

A while back I purchased a handheld barometric-pressure gauge that I could carry with me while guiding and on private fishing excursions. I do not use it every time I am on the water, but I will monitor the device most days. This gauge and my fishing experiences over the past several years have assisted me in determining trends in trout behavior. I am convinced that barometric pressure plays a role in fish behavior and what kind of fishing I can expect. It is important to

Portable barometers can help you monitor atmospheric pressure. Prices range from as little as $30.00 to over $100.00. I have found that you get what you pay for with these devices.

Table 1.1: Barometric Pressure, Weather, and Fishing Conditions

Barometric Pressure	Weather Conditions	Holding Water Where Trout Concentrate
High	Clear skies	Streams—deeper part of water column. Tight to banks and structure. Lakes—drop-offs and tight to vegetation and structure
Transition: Rising	Clearing skies and drying	Streams: Upper part of water column, transitioning to deeper parts as transition persists. Lakes: Upper part of water column, transitioning to deeper parts as transition persists.
Moderate	Fair and stable with some clouds	Inconsistent in terms of where trout are holding and their reactions to patterns and presentation.
Transition: Falling	Cooling temperatures, increasing clouds, chance of precipitation	Very active feeding in a wide variety of holding-water types on both streams and lakes. Potentially the best fishing offered throughout the entire range of barometric pressure.
Low	Cool temperatures, significant cloud cover, precipitation, possibility for thunderstorms	Streams: Trout holding in a variety of water types, but feeding is most consistent below the surface. Lakes: Most activity by trout is occurring below the six-foot level in the water column.

Deep with nymphs on both lakes and streams. Tight to structure and vegetation with streamers on lakes and streams. Tight to structure, banks, seam lines, and the head of riffles with dry flies on streams.

On streams, fish with dry flies until activity subsides, then switch to light streamers and nymphs as transition continues. Target slower holding water and present flies slowly. On lakes, transition from flats and littoral parts of the lake to deeper water as pressure continues to rise. Mimic the natural movement of the trout food you are imitating.

Experimentation is key. Target multiple holding-water types with multiple patterns and presentation on both streams and lakes. Look for consistency.

Streams: Dry flies, nymphs, and streamers in every conceivable type of holding water, especially riffles, seams, flats, and eddies.

Lakes: Streamers and nymphs on shallow flats and upper piece of water column at drop-offs. Heavily weighted patterns on submerged bars and deeper parts of the water column at drop-offs.

Streams: Fish subsurface with nymphs and streamers in all holding-water types. Take advantage of surface feeding when it occurs with surface patterns.

Lakes: Fish streamers and nymphs with movement that matches the natural (or slower) deep on drop-offs, submerged bars, and flats.

understand, however, that a particular river or lake may be affected in a way that is very different from how other waters are influenced.

But even more importantly, each kind of barometric pressure—high, low, moderate, rising, or falling—is accompanied by certain weather tendencies. It could be clear, sunny weather, or it could be substantially cloudy weather. It could be dry, or it could be lots of precipitation. It could be warming significantly, or it could be downright cold. It is easy to get lost in reading a barometer and believing that the fly fishing you are experiencing is directly tied to the pressure of the atmosphere on a given day. What a fly fisher must recognize is that the fishing on a given day can just as easily be influenced by the weather accompanying a phase of barometric pressure.

So while keeping an eye on the barometer when you are fishing, try also to note your observations with trends in the weather. Precipitation, temperatures, wind, sunlight, and cloud cover can have just as telling of a role. They deserve your attention just as much as barometric pressure.

CHAPTER TWO

PRECIPITATION

On a mid-October day in 2010, I awoke to the sound of rain gently tapping the roof of my home in Victor, Idaho. It was 7:00 A.M., and I was scheduled to guide a local real estate agent who had fished with me for the past decade. My drive to Snake River Angler, a fly-fishing shop and guide service in Jackson Hole, Wyoming, takes approximately thirty minutes. As I proceeded down Teton Pass, I could feel the rain intensify. A grin came across my face and the anticipation of what the day might

An approaching storm won't deter this angler on one of Grand Teton National Park's spring creeks. Precipitation can make for uncomfortable fishing, but it can take the bite to a different level.

hold came over me. Cool, wet days in autumn almost always produce outstanding fishing on streams like the Snake River, where we were scheduled to fish that day.

But I knew this guest of mine well, and as soon as I got to the fly shop my cell phone rang with his number on the screen.

"What do you think of this weather, Boots?"

"Well, it's getting me excited, Bill. These are the days we hope for this time of year. It should really increase our chances for a lot of fish and a good size ratio."

"I don't know. I have never had great success in this kind of weather. It just seems to put the fish down. It must disrupt them too much."

Bill was a fair-weather fly fisher, and he was dead wrong about his success in the rain. He and I had fished in this kind of weather at least a half a dozen times over the years, and my log of those days proved that they were well above average in terms of hookups and size. The simple fact of the matter was that Bill was a bigger fan of warm sunshine than he was of clouds, let alone rain.

Despite my encouragement to get out on the water with me, Bill cancelled the trip and switched his reservation to another day later that week. I was lucky enough to get on the river with another guide who was not booked and had the day off. Over the next eight hours, I snapped photos of big, eighteen-inch-plus cutthroats with my smartphone. I texted every one of these photos to Bill.

As a guide and fly fisher, I have learned to be acutely aware of trends in activity on the water. These trends can be in the form of the patterns I am fishing, the presentation I am using, the holding water I am targeting, and the conditions I am fishing in. Statistical reliability comes from having a lot of cases that can be analyzed. Having twenty days that suggest productivity below the surface during phases of high pressure is not nearly as reliable as what two hundred days would suggest.

The trend I see with precipitation is one of greater activity compared to drier and more stable conditions. This is the case whether I am on rivers or on stillwater. I may not be able to target my favorite types of holding water each time or use my favorite patterns and tactics, but the trend definitely shows better fishing.

PRECIPITATION, THE WEATHER, AND FLY FISHING

In chapter 1, I discussed the correlation between precipitation and low barometric pressure. Moist air is less dense than dry air, and moist air is typically part of low-pressure systems and cold fronts. I highlighted my observations regarding the positive influence both low and falling barometric pressure has on fishing productivity. Much of this has to do with the fact that precipitation, which in my experience has a positive influence on fishing productivity, is associated with low barometric pressure.

Precipitation comes in many forms: rain, sleet, hail, and snow, for example. But *levels* of precipitation must be considered as well. Rainy weather can be drizzly, lightly raining, or raining heavily. Snow can be flurries, snow showers, or heavy snow. Each one of these can create different conditions on the water and different trout reactions. If I examine these types of precipitation, my fishing experience has largely been as follows.

Precipitation often accompanies dropping barometric pressure. These two factors work in conjunction to produce the good fishing I encounter when these conditions exist.

DRIZZLE: I define drizzle as precipitation with moisture droplets significantly smaller than that of regular rain droplets. A drizzle itself can be light or heavy, but it is a drizzle because the droplet size is small. I associate a drizzle with falling barometric pressure and, whether I am on a stream or on stillwater, the productivity I note in my log is typically above average. Sometimes it is downright good. In particular, surface action

Many anglers find fishing during periods of precipitation to be productive. A number of holding-water types can have activity, including shallow riffles, seams, and flats. I believe a combination of humidity, dropping atmospheric pressure, and cloud cover work in unison to create ideal conditions.

intensifies on streams with riffles, seams, flats, and eddies producing the most action. Near-surface feeding on lakes—in the top two feet of the water column—intensifies as well. Flats, littoral zones of lakes, and the upper parts of drop-offs produce a lot of action. These are key places in streams and lakes to focus on, but most of the time, a variety of holding-water types will produce. This includes not just those mentioned above, but also on confluence points, troughs, banks, and structure.

LIGHT RAIN: Light rain has well-formed, large droplets with noticeable weight. The frequency of these droplets is less than that of drizzle but more disruptive on stream and lake surfaces because of their size. Like a drizzle, light rain is a feature of falling barometric pressure, but it can also be a feature of stable low pressure or rising pressure (following more substantial precipitation). And like drizzle, my log suggests that light rain generates above-average productivity. Much of the time the same holding water is active as are the same parts of the water column. There can be disturbance of the surface by the droplets on streams and stillwater, but it is generally light enough that it will not disrupt surface feeding and the chances of surface activity if fly fishers are using dry flies or emergers.

MODERATE TO HEAVY RAIN: I define heavy rain as droplets that are (1) oversized, (2) standard-sized but falling at a significant rate, or (3) oversized and falling at a significant rate. Heavy rain is almost always a feature of low barometric pressure. My experience suggests that heavy rain and the low barometric pressure usually does not last long, but occasionally it can stretch on for hours—and sometimes even a day or two. Fishing in moderate to heavy rain can be downright uncomfortable because of the saturation of anglers and gear. But for the intrepid, fishing through periods of heavy rain can also be very good at deeper parts of the water columns. On stillwater, I focus on areas of the water column that are six feet in depth or more. This means I am targeting submerged bars, drop-offs, and flats with substantial depth. On streams, I target deep seams and riffles, as well as deep eddy pools. Six feet may be a bit extreme, but I am fishing closer to six feet than I am to the surface. Of course, the depth I am fishing at is dependent on the depth, velocity, and gradient of the stream.

SLEET: Sleet is the classic "wintry mix" that we hear described by meteorologists across the United States. It is a rain-snow blend that we get as we are transitioning from snow to rain or, more commonly, transitioning from rain to snow. It is caused by snow that partially melts as it descends through the atmosphere. Sleet is generally associated with low barometric pressure, although it can be an aspect of quickly falling pressure. I have not registered many days in my log when sleet figures prominently. What my observations do indicate is that sleet rarely produces noticeable surface activity. The subsurface action, however, can be very good. Like my experiences in heavy rain, I get most of my action in deeper lies in riffles, seams, and eddies, as well as the deeper portions of structure and banks. Stillwater activity is the standard six feet or deeper in the water column. There are times when the action can be higher in the water column on lakes.

HAIL: There is a common misconception that hail is made up of rain droplets that have frozen as they descend to earth. Hailstones are formed in clouds *before* they begin to drop to earth. They are a result of quickly dropping barometric pressure. Hailstorms tend to approach and leave quickly. I cannot recall one that has lasted more than fifteen minutes when I am on the water. During those couple of dozen hailstorms I have been caught in over the years, my fellow fly fishers and I were almost always scrambling for cover. How the fishing is during those times is anyone's guess, and it probably doesn't matter because these storms are quickly moving and are typically over just as soon as they begin.

FLURRIES: Snow flurries are characterized by light snow that falls in an intermittent fashion. The catch is that this intermittent light snow can occur off-and-on throughout a day. Flurries are associated with falling or low barometric pressure. Air temperatures are generally cool if not downright cold. During days when flurries dominate the weather scene, my experience is that fishing is generally good with surface patterns when emergences are occurring. Such conditions on Rocky Mountain streams produce hatches of chironomids and blue-winged olives, as well as PMDs (pale morning duns) from time to time. However, the action can be even better below the surface with nymphs and streamers. This is especially the case on stillwater. My retrieval speeds are often slow with a standard streamer retrieve or a figure-eight/hand-twist retrieve. Most of

This late October day on Lewis Lake in Yellowstone National Park was filled with cloud cover and light, but steady, snow flurries. We fished streamers over flats and drop-offs, with slow retrievals. The bite was happening the entire time we were on the water.

the activity I observe occurs on submerged bars, drop-offs, and on the deeper portions of flats. On drop-offs and flats, action is typically at or below the four-foot level of the water column.

On streams, I get action in a variety of water types (everything from riffles to structure to seams and eddies) and at a variety of depths. Dead drifting nymphs is my standard tactic, but I also get action when I swing or jig them through riffles, seams, flats, troughs, and confluence points. When fishing streamers, I vary my retrieval speed and length (amount of line brought in on each strip) until I find consistent action with a particular presentation.

Surface action on streams can also be in the cards when there are flurries. I recall one October day on the Madison River, where snow flurries dominated the weather. Midges, minute caddis, and blue-wing olives were emerging throughout the afternoon, and I had exceptional fishing on #18 olive

This early March storm brought intermittent snow showers to the South Fork of the Snake River in Idaho. I typically fish below the surface with double- or triple-nymph rigs and streamers in such conditions.

comparaduns and black Furimsky BDEs. Most riffles and seam lines were productive. This is the case on most streams I fish when flurries are in the forecast.

SNOW SHOWERS: Snow showers are light snow that falls periodically or throughout an entire day. Like flurries, snow showers are often associated with falling or low barometric pressure. My experience suggests that there can be limited surface activity on riffles and flats, but there is more productivity below the surface with nymphs and streamers in a variety of water types. I will generally dead drift my nymph rigs, but there can be some action on nymphs when they are swung or slowly retrieved. My observations when fishing on stillwater during snow showers is similar to what I experience during flurries.

I have more success at deeper parts of the water column. On flats and drop-offs, I am usually fishing at the four-foot level or deeper with both streamers and nymphs, employing slow to moderate retrieval speeds.

MODERATE TO HEAVY SNOWFALL: Where I live in the Rocky Mountain West, this is the weather that most of us who ski or snowboard dream of. Experienced fly fishers will tell you that these conditions can produce worthwhile results on many streams and lakes.

I can distinctly remember an early October excursion I took to Lewis Lake in Yellowstone National Park several years back. Our region was hit by an early autumn storm that pro-duced significant snowfall above 7,000 feet elevation. That day, Will Dornan and I fished the mouth of the Lewis River Channel, the drop-offs extend-ing out from the Brookie Bay Flat, and a lesser-known feature called Mackinaw Point with sinking lines and moderately sized streamers. From mid-morning until we motored back to the boat launch just before dusk, we experienced heavy snow. It was wet and uncom-fortable, but the fishing was out-of-sight. I was expecting a good day, but not as blockbuster as it turned out to be.

Moderate to heavy snow is snowfall that lasts for at least two hours in duration. Precipi-tation droplets—in the form of

Heavy snowfall hits the Teton River in early April. While going subsurface is the general rule in such conditions, it is possible to get heavy emergences of midges and blue-winged olives. When this occurs, and fish are actively feeding on the surface, I will make a quick switch over to a dry fly or tandem dry rig. Fishing can be uncomfortable, but the results are often worth it.

snowflakes—are generally thick and can contain significant amounts of moisture. Such snowfall is correlated with falling and low barometric pressure.

As my brief story from Lewis Lake suggests, fishing during periods of moderate to heavy snowfall is best below the surface. Sometimes you must focus your attention well below the surface. On stillwater, I target the deeper part of the water column (six feet and deeper) on drop-offs. This is where I experience most of my activity. A close second in terms of activity is on submerged bars. These bars can be difficult to reach with fly-fishing gear, especially when they are twenty-plus feet in depth. Flats can also produce if there is sufficient depth. I retrieve my streamers and nymphs in a way that matches the natural I am imitating. Streamers tend to produce better than nymphs.

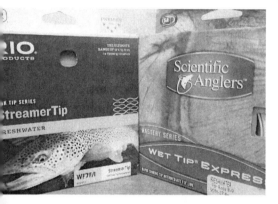

Sinking lines allow fly fishers to target deeper portions of stillwater—six feet or more— effectively. This is where a lot of the action can be when moderate to heavy snowfall is occurring. Intimate knowledge of your line's sink rates will help you target this part of the water more efficiently.

When fishing streams during periods of moderate to heavy snowfall, I am almost always below the surface, and a good part of the time it is with streamers. I will fish streamers with moderate to fast retrieval speeds through riffles, troughs, confluence points, seams, and along banks and structure. I am almost always using a sinking tip or line that is a Type III or greater, depending on the depth and gradient of the particular stream. Nymphing can be productive, but I am usually getting action in deep riffles and seams and, at times, on flats that are three feet in depth or more.

Keep in mind that moderate to heavy snowfall can at times produce emergences on those insects that we typically think of when the weather is cold and wet. Chironomids and blue-winged olives are what typically come to mind. I observe these occurrences in early spring on the waters I fish. So while I may be subsurface, I will certainly make the switch to a dry fly or tandem dry rig when fish are feeding on the surface.

"RAIN MAKES THE WATER COME ALIVE!"

It is easy to tell from this chapter thus far that I am a big fan of precipitation. My experiences tell me that, most of the time, rivers and lakes get an added punch of activity when it is raining or snowing. One of my favorite clients over the past two decades is also a fan of precipitation. He always tries to book with me in early spring and late autumn when the chance for rain or snow, as well as cooler air temperatures, is better than other times of the year. When precipitation starts to pick up, my friend

Some anglers who target sea-run fish believe that precipitation alters the chemical composition of rivers. It is this change that sparks the much anticipated spawning runs upstream.

Table 2.1: Precipitation, the Weather, and Fly-Fishing Strategies and Tactics

Precipitation	Barometric Pressure	Holding Water Where Trout Concentrate
Drizzle	Falling to low	**Streams:** Riffles, seams, flats, and eddies. Banks, structure, and confluence points at times. Upper portion of water column. **Lakes:** Flats, littoral zones, and drop-offs. Upper portion (top two-to-three feet) of water column.
Light Rain	Generally falling or rising, sometimes low	**Streams:** Upper part of water column in riffles, seams, flats, and eddies. **Lakes:** Similar to periods of drizzle; upper part of water column on flats, littoral zones, and drop-offs. Deeper portions of water column—down to six feet—can be productive at times.
Moderate to Heavy Rain	Generally low	**Streams:** Deeper portions of water column in riffles, seams, and eddies. **Lakes:** Deeper portions of water column (six feet or deeper) on submerged bars, drop-offs, and flats.
Sleet	Generally low, sometimes quickly falling	**Streams:** Deeper portions of water column in riffles, seams, and eddies, as well as along structure and banks. **Lakes:** Deeper portions of water column (six feet or deeper) on submerged bars, drop-offs, and flats.
Flurries	Falling or low	**Streams:** A variety of holding-water types at a variety of depths within the water column. **Lakes:** A variety of holding-water types at and below the four-foot level of the water column.
Snow Showers	Generally low, sometimes falling	**Streams:** A variety of holding-water types primarily at deeper portions of the water column. Some activity at the surface on riffles and flats. **Lakes:** Deeper portions of the water column, primarily on flats and drop-offs.
Moderate to Heavy Snow	Low	**Streams:** Deep in a variety of holding-water types. **Lakes:** Six feet or deeper in the water column on drop-offs and submerged bar. Flats can hold trout near the lake bed if there is sufficient depth.

Streams: Dry flies and lightly- to moderately-weighted nymphs in most holding-water types, but especially riffles, seams, eddies, and flats.

Lakes: Two-to-four-foot level of the water column of flats, littoral zones, and drop-offs with nymphs and streamers. Floating line or intermediate tip or line.

Streams: Dry flies and light- to moderate-weight nymphs in most holding-water types. Switching to nymphs if surface action subsides.

Lakes: Nymphs and streamers on floating or intermediate tips and line at upper portion of water column from surface down to six feet. Target flats, drop-offs, and littoral zones.

Streams: Double- or triple-nymph rigs with substantial weight and leader length. Streamers with substantial weight or heavy tips or sinking lines.

Lakes: Heavy sinking lines for streamers and long leaders for nymph rigs, both of which will allow flies to reach the ideal depth at which trout are holding.

Streams: Double- or triple-nymph rigs with substantial weight and leader length. Streamers with substantial weight or heavy tips or sinking lines. Targeting a variety of holding-water types.

Lakes: Heavy sinking lines for streamers and long leaders for nymph rigs, both of which will allow flies to reach the ideal depth at which trout are holding.

Streams: Nymphs and streamers in a variety of holding-water types and portions of the water column with dead drifts, swings, and jigging. Dry flies and emergers in riffles, seams, flats, and eddies.

Lakes: Nymphs and streamers fished at or below the four-foot level of the water column on flats and drop-offs. Slow standard retrieve and figue-eight/hand-twist retrieve.

Streams: Dead-drifting, slow-retrieving, or slow-swinging nymphs in a variety of holding-water types. Slow-retrieve and slow-swinging streamers in a variety of holding-water. Surface patterns at times in riffles and flats.

Lakes: Nymph rigs with sufficiently long leader and streamers, as well as heavy, sinking tips and lines to reach the four-foot depth or deeper in the water column on flats and drop-offs.

Streams: Streamers fished with Type III or heavier sinking tips or lines, depending on stream gradient and velocity. Moderate to fast retrieval speeds. Double- or triple-nymph rigs in riffles, seams, and confluence points at three feet or deeper on most streams.

Lakes: Streamers with heavy, sinking lines or nymphs rigs with sufficiently long leaders to allow flies to reach the six-foot level of the water column or deeper. Primarily drop-offs and submerged bars.

gets downright giddy and exclaims "rain makes the water come alive!"

Rain does indeed make the water come alive. That is what I have typically observed over my lifetime as a fly fisher. Why is this so? There are several theories advocated by experienced fly fishers. One of these has to do with cloud cover and the belief that trout feed more aggressively under cloudy skies. Basically, it is cloud cover, not precipitation, that is the key. This premise, however, doesn't actually answer the question, and if anything creates another question: why is trout feeding behavior altered with cloud cover?

Another assumption is that precipitation alters the chemical composition and pH level of streams and lakes in a way that produces more activity on the part of trout and the food they consume. Several European fly fishers expressed this belief to me when I was guiding in Tierra del Fuego, Argentina. January rains would raise the levels of the region's Rio Grande and Rio Menendez, and after the rivers receded, the fishing activity would increase significantly. Of course, we were fishing for sea-run brown trout, and my experience with anadromous game fish like steelhead and salmon was that rising stream levels allowed these fish to continue their migration to pools and runs farther upstream. It was this activity, coupled with their spawn-induced aggression, that most likely caused more intense action. A change in pH probably had very little to do with it.

These guests from across the pond, however, claimed that precipitation and the resulting change to a river's pH had the same impact on stream-resident trout almost everywhere they had fished around the world. This included everything from brown trout in New Zealand to rainbow trout in Oregon.

It is an interesting proposition. The impact of pH on trout water cannot be understated (it is something that I examine in a later chapter). But I believe a more plausible reason for the very noticeable increase in activity I experience has to do with the effect precipitation has on aquatic insects, particularly mayflies.

Many fly fishers observe emergences, sometimes prolific, of various mayfly species during days or periods of a day with precipitation. This phenomenon is well documented in contemporary fly-fishing literature. In *Western Mayfly Hatches*, authors Rick Hafele and Dave Hughes write about the emergence of blue-winged olives on wet and even snowy days. Malcolm Knopp and Robert Cormier in *Mayflies* note the same about numerous species of the order Ephemeroptera.

My own experiences suggest the same. One of my favorite autumn hatches in the Yellowstone region, the mahogany dun, is most prolific during cloudy and rainy days. A drizzly, mid-October day on Montana's Bitterroot River stands out. While fishing the river just upstream of the town of Lolo, I was able to hook into several trout during a three-hour period with unweighted Pheasant Tail Nymphs and Halo Emergers in riffle after riffle. It remains the best day I have had on the Bitterroot in the half dozen times I have fished it.

Dehydration is the most common killer of mayflies after emergence. This may be the reason why we see prolific hatches when it rains or snows. Precipitation gives them a greater chance to extend life and mate successfully.

Another instance that stands out in my mind is a late June day in 2004 on the South Fork of the Snake River in Idaho. My friends and I put on the water specifically to fish the salmon fly hatch, which was making its way slowly up the Lower Canyon reach. The morning and first part of the afternoon were active with emerging salmon flies and we were picking up trout in all the typical pieces of holding water one would expect during a giant stonefly emergence, particularly along banks, structure, and flats. But as the day progressed, dark clouds from the west moved up the river and with them came precipitation that moved back and forth from light to moderate rain. There were still some salmon flies about, but their significance on the water was replaced by an impressive emergence of pale morning duns. Trout began to feed exclusively on these mayflies in riffles, seams, eddies, and flats. We focused our attention in these holding-water types, and when we eventually reached our takeout, my companions and I were in agreement that fishing during the afternoon's

PMD emergence completely outperformed the fishing earlier that morning when salmon flies dominated the scene.

It is not just blue-winged olives, mahogany duns, and pale morning duns that emerge during periods of precipitation. I have also witnessed hatches of green and brown drakes, flavs, tricos, *Callibaetis*, *hecubas*, and March browns. This is not to say that these mayflies do not hatch under sunny skies and warm weather. But when the clouds roll in and the rain or snow comes down, I am almost certain to see mayflies on most of the waters I fish for trout.

Why does this phenomenon occur? Why do hatches of various species of mayflies tend to intensify when there is precipitation? The answer most likely lies not in how mayflies live, but rather how they die. What we know about mayflies is that, when they die naturally, it is from dehydration. Many experienced anglers believe that mayflies receive an instinctual impulse to emerge during periods of cool, cloudy weather with precipitation because they will have a chance of living longer during their adult stage, which in turn gives them a better chance at mating.

This is a significant reason why I experience solid surface action during cloudy weather with a drizzle, light rain, snow flurries, and snow showers. The water I am targeting during this weather is where I would typically find mayfly larva and emergers—riffles, flats, seams, and eddies. If I am using nymphs or streamers, I am often fishing high in the water column or in relatively shallow water. This is where emerging mayflies are at their most vulnerable, where egg-laying females will be available, and where spent spinners can be found by trout.

When the rain comes down, mayfly emergers are my pattern of choice. Quigley Cripples, Film Critics, and Booty's PMD Emerger (pictured here) are among my favorites.

CHAPTER THREE

WATER TEMPERATURE

I remember the first time I fished the fabled San Juan River in New Mexico. It was early May 2001, and I was returning home to Wyoming from my first year of graduate school at the University of Texas in Austin. My carefully planned road trip took me to the base of the dam at Navajo Lakes, where one of the best tailwater fisheries in the world begins.

It was 80 degrees when I arrived. This was unseasonably warm, even for this part of the Land of Enchantment. As I prepped my gear, I witnessed two fly fishers decked out in full chest waders making their way down the embankment from the parking lot to the river. *You got to be kidding me*, I thought to myself. Way too warm for that kind of attire. My first assumption was that these guys, easily in their sixties, had lost their ability to adapt to cool water temperatures in their later years. I was a young buck. There was no way in hell I was going to dress up in waders when it was that warm.

Wearing shorts, water shoes, and a short-sleeve shirt, I made my way down to the water. I stepped in to make my first cast and lasted about thirty seconds before exiting for dry land. In the research I had done before that road trip, I had taken everything into consideration except for one very basic factor—water temperature. It was *43 degrees!* I immediately retreated back to my truck and jumped into my waders. Several laughs went up from others in the parking lot, and one fellow fly fisher told me that this was not the first time that he had seen a San Juan

Rookie have to come back and put on the right kind of apparel. Totally embarrassing rookie move indeed.

I have fished the San Juan a few times since then. No matter what the weather is, I have been in waders every time.

Air and water temperatures can be widely independent of each other. There are many streams and lakes where air temperatures have a very noticeable influence on the temperature of water. Air temperature can also have an impact on the availability of certain types of trout forage. Regardless of the influence, water temperature trumps air temperature in importance. It is unquestionably the most important component of a trout's world.

WATER TEMPERATURE AND TROUT TOLERANCES

In *Fish Hatchery Management,* author Gary Wedemeyer states that no other single factor affects the development and growth of fish as much as water temperature. Metabolic rates, feeding activity, spawning, and egg hatching are all influenced by the temperature of the water in which fish live. Different trout species have ideal temperatures with extremes at both the high and low ends of the scale. Char, which include all species of the genus *Salvelinus,* differ from other trout in their adaptation for cold water. The optimum range for life-sustaining activity is considered to be between 50 degrees and 57 degrees. Even among the six different char species there is wide variation. Brook trout, for example, have a greater tolerance for warm temperatures than their char cousins. Lake trout, on the other hand, can thrive at temperatures well below the 50 to 57 degree optimum range. Another char, the arctic char, has even more extreme adaptations. Robert Behnke, in his renowned book *Trout and Salmon of North America,* notes that members of this species have been documented feeding at temperatures near the freezing point (Behnke suggests that there may be anti-freeze-like substances in their blood, similar to that found in other arctic marine fishes).

Now take trout of the genus *Onchorhynchus,* which includes rainbow, cutthroat, and golden trout. These three forms of trout can be separated into *at least* forty different subspecies. Each of these subspecies has unique differences in terms of habitat and forage preferences. However, the general thermal tolerance range for fish of this genus is between 32 degrees and 77 degrees, with an optimum feeding range of between 50 degrees and 65 degrees. Remember that these are just generalities. There are more than enough documented cases of rainbows, cutthroats, and golden trout feeding above, and well below, this optimum range. In fact, the Bear Lake strain of Bonneville cutthroat display an increase in weight during winter months when water temperatures are well below 40 degrees (much of this has to do with their feeding on cisco minnows, a species of whitefish, which spawn in January and February).

Brown trout have been stocked the world over from their native waters in Europe for a good reason—they are highly adaptable. They can withstand higher water temperatures (sometimes up to the low 80s) and can feed at higher water temperatures than rainbow and cutthroat trout. This makes them well suited for waters that other trout may not be able to handle. The Firehole River in Yellowstone National Park is a good example. It flows through a basin filled with thermal features. Geysers and hot pots spill hot water into the stream along its entire length. Previously fishless, the Firehole was stocked with brown trout and rainbows in the 1890s.

Table 3.1: Optimum Temperature for Activity on Streams and Stillwater by Trout Species

Trout Species	Optimum Temperature for Activity on Streams	Optimum Temperature for Activity on Stillwater
Rainbow	55 degrees to 66 degrees	59 degrees to 68 degrees
Brown	50 degrees to 68 degrees	59 degrees to 68 degrees
Cutthroat	53 degrees to 66 degrees	58 degrees to 66 degrees
Brook	52 degrees to 61 degrees	56 degrees to 62 degrees
Lake	N/A	46 degrees to 60 degrees

I guide on the Firehole River a few times each year. Most of my focus is on the early part of the season from Memorial Day weekend (when Yellowstone opens to fishing) until the middle of June, and then again in October. Water temperatures are usually ideal at these times of the year. Every now and then, however, I will take a party out in July when water temps can be well into the 60s and sometimes boarding on 70 degrees. When this occurs, the best plan is to fish the cooler tributaries like the Little Firehole and Iron Springs Creek. These waters are where many trout, especially the larger ones, will retreat to when temps on the Firehole become uncomfortably warm. But this also renders them crowded. If there are just too many fly fishers, my party and I will fish the Firehole. Despite the very warm temperatures, fishing can be decent because the brown trout are still active.

Table 3.1 displays the optimum water temperatures for feeding activity on streams and stillwater amongst the trout species targeted by most fly fishers in North America. Please keep in mind that what is optimum can vary from one stream or lake to another. Lake trout on Jackson Lake in Wyoming can react differently to water temperatures from their cousins on Lake Pend Orielle in the Idaho Panhandle. Subspecies of trout matter as well. I can tell you from experience that the Bear River strain of Bonneville cutthroats have a wider threshold of temperature tolerance than any other subspecies of cutthroats I have fished for.

WATER TEMPERATURES AND TROUT WATER COMPOSITION

A good number of us carry a thermometer when we are on the water. I have been doing this for the last fifteen years. It does not always make it into the water I am fishing, but I do find it handy for gauging differences in activity over a day of fishing. When I do use it, I try to document at least two temperature measurements. Usually the first one is taken midmorning and the last is taken just prior to finishing up the day. There are times when I will make midday measurements as well. This data allows me to observe three important variables connected to water temperature:

1. Fluctuations in a given day

2. At what temperature trout begin or cease to feed on the surface and below

3. At what temperature feeding is most active

How dramatic temperature fluctuations can be on a given stream is dependent on various factors. Air temperatures and spring upwelling (be they hot springs or cold springs) can play a role. There are also factors such as whether a stream is a tailwater or has more freestone characteristics. Generally, streams that are defined as spring creeks or tailwaters have cool, consistent water temperatures, while freestone streams can have temperatures that are much more in flux throughout the year, or even a given day. But do not take these things to be written in stone. Some very popular tailwaters are developed for irrigation. The reservoirs feeding them can be drained significantly in mid to late summer. When this occurs, the low water remaining in the reservoir can experience temperature spikes well into the sixties and sometimes even the low seventies. This water is then fed into the downstream tailwater.

Digital thermometers allow for a quick and accurate reading of surface water temperatures. I still like the old-school stream thermometers. I attach mine to a piece of color-coded Dacron separated into two-foot sections. This allows me to submerge my thermometer to the depth of the water column that I want to measure.

Conversely, not all freestone streams experience extreme fluctuations in their water temperatures throughout the year. Some freestone streams have serious coldwater tributaries and spring-water upwelling that can keep them cool even when air temperatures are very high.

WATER TEMPERATURE AND TROUT HOLDING WATER

Water temperature controls all in a trout's world. It dictates their metabolism, when they spawn, when their eggs hatch, when they eat, what they eat, how much they eat, if they live, and if they die.

Consider the role water temperature plays in the availability and variety of forage available to trout. Aquatic insects hatch in narrow temperature ranges. During those cold months of winter on most streams in North America, aquatic invertebrate emergences are generally confined to chironomids, blue-winged olives, and winter stoneflies like those of the genus *Capnia* and *Nemoura*. These are the ultimate coldwater bugs and the reason they are so prominent for fly fishers from December through March.

Chironomids are the one true constant on trout streams no matter how cold water temperatures become. Midge imitations might be the only game in town on those cold midwinter days.

As air and water temperatures warm in the spring, pale morning duns and a multitude of caddis species begin to make their appearances. As spring moves into summer and water temperatures continue to warm, we experience a literal onslaught of aquatic insects like drakes, salmon flies, golden stones, little yellow and little green stoneflies, *Callibaetis*, dragonflies, and damselflies.

When fall arrives, blue-winged olives and midges become more prominent again. They are joined by October caddis and mahogany duns. There are variations from stream to stream and lake to lake, but we are seeing these invertebrates in large part because their emergences and activity are dictated by changes in water temperature.

More important than food availability is where trout hold and when they feed.

Trout are poikilothermic creatures, meaning that their body temperature varies with the temperature of the surrounding environment.

Whatever the water temperature is in a stream or a lake, the body temperature of trout is most likely going to be within a couple degrees of that water. This characteristic greatly impacts their activity.

When water temperatures are extremely low—say in the 30s—there is little activity among most aquatic organisms. Many invertebrates go into a dormant or semidormant phase. Like trout, their body temperatures are also tied to the temperature of their environment. Some of the few invertebrates that are active include chironomids, tiny winter stoneflies,

Warming water temperatures in early summer spark the emergence of a rich array of aquatic insects. Salmonflies hatch when temperatures hit the mid to upper 50s. They can be found on famed rivers in the western US from May to July each year.

and crustaceans like scuds and shrimp. There is food available, but there is not much of it. Generally trout will feed when forage is near their holding position. Any movement toward prey that is not in their immediate proximity can result in the expenditure of energy that cannot be replaced by the available forage. This is the case for trout in both streams and stillwater. These coldwater conditions often lead to minimal growth for trout. They can lose weight. Extended periods of extremely cold water temperatures can lead to high mortality among trout populations, particularly on streams. This is especially the case for young trout—less than two years of age for rainbow and cutthroats, and less than four years for brown trout.

However, while there is less forage available, there is also less need for energy intake by trout because of their slower metabolisms. Jason Randall points out in his book *Feeding Time* that trout raised in hatcheries are fed one-quarter of the amount at 40 degrees that they are at 60 degrees. Fish also benefit from the ability of cold water to retain oxygen. This allows them to revive faster during times of stress, such as fleeing predators, chasing prey, or being hooked by a fly fisher.

As water temperatures warm into the 40s and low 50s, more forage is available to trout in the form of chironomids; *Baetis*; winter stoneflies;

Warm water temperatures in August are a time of stress for trout on a number of streams across the country. Tailwaters like the Green River (pictured here) and San Juan River are the exception. The deep reservoirs feeding them keep stream temperatures ideally regulated.

Skwalas; pale morning duns; and, seasonally dependent, mahogany duns; and October caddis. The warming temperatures allow the body temperatures of trout to warm as well. Again, this is the case for trout residing in both streams and stillwater. Lethargy is less of an issue and trout can move more easily. This is where energy loss typically subsides and trout begin to achieve gains in size again.

The low 50s to mid-60s is the time of bounty for trout. Their metabolism is peaking. There can be a lot of energy expenditure, but forage is maximized. Most of the trout foods available at lower water temperatures remain available. Added to the mix are invertebrates that emerge at warmer water temperatures. These include large insects like drakes, golden stoneflies, salmonflies, crane flies, dragonflies, and damselflies. Trout can handle positioning themselves in faster, energy-burning currents because there is more forage available.

When water temperatures begin to increase to the high 60s and into the 70s, an intensely stressful time begins for trout. There can still be a lot of forage available in the form of aquatic invertebrate larvae and pupae. Because warm water temperatures often correlate with warm air temperatures, terrestrials are typically available as well. These warm water temperatures, however, lead to increased body temperatures for trout and, thus, a higher metabolism. Warm water also contains far less oxygen than cool water. It is a double whammy for fish in a serious, life-threatening way.

Trout cope with this chaotic environment through the process of thermoregulation. With the inability to regulate their body temperatures internally, trout will use their surroundings to raise and lower their internal temperature. In essence, they are using external mechanisms to control their body temperature. If the temperature in a particular piece of holding water is too warm, they will move to another part of the stream or lake that is more hospitable.

A number of factors can cause temperature variations on streams and lakes. Perhaps nothing is as important in this regard as depth. Temperatures are often stratified in a column of water. Typically, warmer water is at the top of the column and cooler water is at the bottom. This is due to the inability of the sun's rays to penetrate the deeper portions of a stream or lake. I have measured the temperature of the water column on streams like the Madison when the surface came in at 66 degrees and the temperature at the four-foot level was 62 degrees. This does not seem like much, but it means a lot to trout. The difference on lakes can be dramatic. I recall fishing Montana's Hebgen Lake in late June one year when the surface temperature was at 64 degrees and the six-foot level came in at between 57 and 58 degrees.

This Hebgen Lake rainbow was caught on a very warm summer day at the six-foot level in the water column.

The lakes I fish in July and August reach temperatures that render the top six to ten feet of the water column barren of feeding trout most of the day. I head deep with lines and tips ranging from slow-sinking hover and intermediate lines to eight-inch-per-second sinking tips. Intimate knowledge of your sink rates is important in these conditions.

Keep in mind that this is just a rule of thumb. Some natural phenomena create different conditions. Stillwater is a good example. When lakes go into ice-out in spring, the water at the surface is close to 32 degrees while the temperature at the bottom is closer to 40. These conditions will remain until lakes turn over. At this point, the surface temperatures will be warmer than the water at the bottom of a lake. This process will reverse itself in autumn when many lakes turn over once again.

As water temperatures close in on the mid to high 60s, trout on rivers will often descend to deeper pieces of holding water, especially the deeper portions of riffles, seams, and eddies. Flats can be barren unless there is enough depth to create comfortable temperatures, a very real possibility on streams with a low or moderate gradient. I live in serious dry fly country, but I have no problem throwing nymphs. I actually enjoy fishing nymphs, an attitude that is still sacrilege for the dwindling number of dry fly purists on my home waters. When I am faced with these conditions, I go deep with double- or triple-nymph rigs. I am looking for strict dead drifts with a little movement created at the end of my presentation with swings and rod lifts. When fishing lakes in warm-water conditions, movement is the name of the game. I focus on retrieves and rod lifts that match the movement of invertebrates I am imitating. Drop-offs and deeper flats (with depths of eight feet or more) are my primary targets.

Streamers can come into play when fishing in warm-water conditions. On streams, I typically fish with tips that have a sink rate of between three and six inches per second. This is all dependent on the depth and current velocity. I am not married to any one particular retrieval speed when fishing streamers deep in warm-water conditions. My observations over the past three decades do not suggest that one speed outperforms others.

On lakes, my sinking lines or tips usually fall somewhere between a slow-sinking hover line and a Type IV line similar to Rio's Deep Lake line. My retrieval speeds are slower with a hover or intermediate line than with a faster-sinking line. Once I have my streamer down in the feeding zone, I will speed up my retrieves. Focusing on your sink rate on lakes is important. Allowing your streamer to get down to where trout are feeding might be the best thing fly fishers can do to get hookups.

The attraction of a large dry fly should not be ignored in warm-water conditions. Trout might be holding in deep lies, but this does not mean that a sizeable form of potential energy won't draw them to the surface. I have seen this numerous times on freestone-like streams in the Rocky Mountain West, particularly on streams in my home state of Wyoming such as the Wind River, Snake River, and the upper Green River. These streams can get late-summer hatches of stoneflies like short-wing stones that can grow to two inches in length. Just as important are big terrestrials: grasshoppers, beetles, and ants. There are those days when surface action on #12 or smaller mayflies and caddis imitations comes to a screeching halt after 1:00 P.M. Switching over to #10 or #6 dry attractors can keep the surface action going.

Getting trout to the surface with big dries is one thing. Getting

Thermoregulation allows trout to regulate their body temperature. They hold at the mouth of springs and deeper in the water column to keep their bodies cool when water temperatures are warm.

them to strike is another. Fish may get close to your offering, but they must break through an uncomfortably warm layer of water near the surface. Many anglers take this to be a refusal. However, astute fly fishers believe that this "refusal" is actually the trout reacting to that warm two or three inches at the top of the water column. I will remedy this situation by using a short dropper nymph suspended fourteen to eighteen inches below the dry fly. On a three-to-five-second drift in a current of moderate speed, the nymph can descend between five and ten inches on average. This will get it below that warm top layer. While the fish may be deflected by the warm surface layer of water and not take the dry, they still have a very good chance of taking the nymph

Other factors are worth considering when water temperatures are warm. One is the presence of tributaries or springs. These can discharge cool water with tolerable temperatures for trout. I attended a presentation on redband rainbow trout survival mechanisms that rammed this point home for me. One of the presentation's photographs showed two redbands between six and eight inches holding at the mouth of a tiny spring while the rest of the stream within the frame of the photograph was devoid of fish. A small icon near the spring displayed a temperature of 25 degrees Celsius (77°F). Another icon set in the middle of the stream showed a temperature of 27 degrees Celsius (81°F). That little piece of holding water matters. It is crucial to their survival, especially when water temperatures are that high.

Another aspect to consider is time of day. During those times of the year when water temperatures are peaking, the best action is often at and just after dawn. This is when water temperatures are cool due to cold overnight air temperatures and a lack of radiant heat during nighttime hours. This has been the case on many of the waters I guided on over the years. The year 2007 especially stands out for the drought in the Rocky Mountain West. Most drainages received less than 60 percent of their average snowpack. Added to this was paltry precipitation during the spring. Fishing was terrific from April until the middle of July. The hot temperatures that summer led to water temperatures that exceeded 64 degrees Fahrenheit on most streams that I guided on that year. Fishing wasn't just slow during the afternoon, there was serious stress put on trout once they were hooked and eventually landed. Some very concerned anglers would purposely break their fish off once they were hooked, an act that decreased the chance of mortality. To remedy matters for both guests and trout, we began meeting our parties early to start the day. Instead of the usual 8:00 A.M. rendezvous at the shop, we would meet at 6:30 A.M. Our trips would finish up around 3:00 P.M. instead of the typical 5:00 P.M. to 6:00 P.M. The fishing was excellent from the time we put on until just after noon. As an added bonus, we were off the river before the stifling heat of late afternoon.

Such conditions are at work on stillwater as well. In fact, they probably play more of a role. Early in the season—May through early July generally—most of the lakes I fish are as easy as pie. You put on between 8:00 A.M. and 9:00 A.M. and fish until late afternoon with terrific results. But as the summer drags through July and into September, the sun's rays heat the standing water well into the high 60s and sometimes the low 70s. When this occurs, success with a fly rod during the "normal" hours is tough unless one is going deep. I compensate by starting the day early,

This twenty-inch Snake River cutthroat was caught amongst a pod of fish on an early March day. Water temperatures were at 39 degrees in midafternoon. When you find a pod, stay on it.

6:00 A.M. to 6:30 A.M. It seems like simple advice—and it is. As the year drifts into late autumn, conditions tend to swing back to mid- and late-day productivity.

Many anglers find extremely cold water conditions to be more difficult to deal with than warm water. Not only is it uncomfortable, productivity can seem futile unless we tamp down our expectations. Cold water means slow metabolism. It can take a lot of effort to get hookups. Trout will often descend to deeper holding water where water temperatures start to hit 40-degree levels or lower. Part of this is due to there being slightly warmer temperatures in deeper water. But the key here is that these temperatures are *slightly* warmer, and sometimes the difference is negligible. More importantly, trout are holding in deeper portions of the water column because this is where currents are slowest. With a slower metabolism and limited forage, survival is better guaranteed in these portions of a stream.

Cold-water conditions generally occur during winter months when most streams are at their lowest in terms of flow and volume. Habitat is reduced. Some holding water dries up and disappears. The advantage here is to the fly fisher. Trout are forced to congregate in available holding water. This act is referred to as *podding up*. Trout are holding in pods in specific parts of a stream. When you find a pod and fish are feeding, it is possible to get into several hookups despite frigid water.

I am lucky enough to guide year-round in the Snake River region of Wyoming and Idaho. This includes the winter months from December through March, when our area streams are at their coldest. I have a diverse set of streams where I can take guests. This includes high-gradient streams like the Snake River in Wyoming, streams with moderate gradient like the South Fork of the Snake River in Idaho, and low-gradient streams like the Salt River in Wyoming. I get to see where trout pod up every winter.

So where do trout pod up during cold-water conditions? The two prime pieces of holding water where I find these pods are riffle pools and seams, both with sufficient depth.

On high- and moderate-gradient streams, I tend to find pods of trout holding at the middle and tails of riffle pools. Low-gradient streams can have pods holding throughout the riffle pool. Trout will be deep in the center of the riffle pool and at the current margins. Regardless, I try to fish throughout the pool from head to tail using double- or triple-nymph rigs with sufficient weight to get the rigging down where trout are holding. Sometimes I use external weight in the form of split shot, but most often I am using weighted nymphs. It all depends on the depth of the water column I want to reach and the speed of the current running through the pool.

Trout do not move too far from their holding position when water temperatures

Fish will not move very far to your offering if water temperatures are 40 degrees or lower. Dissecting your target into a set of grids and covering each part methodically will help assure a better chance for hooking up.

are below 40 degrees. The fly fisher needs to make several passes through the same area of holding water and cover every part of it. I divide the holding water I am targeting into a grid pattern. Each square within the grid is three to five square feet. I will make four or more casts from a position before moving and casting again. With each cast, I will be covering a different horizontal piece of my target. The first set of casts are in rather close—within ten feet of my position. This is similar to short-line and tight-line tactics I described in *Modern Trout Fishing*. The next set will be farther away from my position by two to three feet. The next set will be yet another two to three feet away, and so forth. Coverage is crucial. The idea is to hit every conceivable square within the grid with at least four drifts before moving on to the next square. Sometimes, you may require a half dozen or more drifts through a particular piece of holding water before a take occurs.

I tend to fish with no more than twenty feet of line and leader extended from my rod tip in cold temperatures. Shorter lines are easier to cast in these conditions, and you are less likely to deal with problems of ice buildup on the line and guides. In addition, takes in colder water are often subtle. Seeing and feeling a strike is easier with a shorter line.

Fishing seams are interesting in cold-water conditions. Many seams can be devoid of fish because stream flows are at their minimum. There is not enough depth to sustain even small pods. But the one type of seam that tends to hold trout in high numbers during cold-water conditions are those that run alongside channels with minimal or no current. I refer to these as eddy seams. These seams are characterized by a standing or near standing pool that flanks a faster-moving current. There is often a slower current at the tail of the seam that recirculates back upstream but dissipates well before it meets up with the seam again.

In warm-water conditions, trout often hold on the current line separating the main channel from the standing pool or back eddy. Metabolism rates are high enough to hold here. More importantly, warmer water means available forage. In cold-water conditions, the low metabolism of trout will keep them from holding on this current line. They will instead hold in the standing pool where currents are minimal. I have come upon these pools while wade-fishing and observed scores of trout suspending in these eddies. I target these trout with lightly weighted double- or triple-nymph rigs imitating swimming larvae like blue-winged olives, *Callibaetis*, gray drakes, damselflies, and dragonflies. A good rule of thumb is to fish these without an indicator and retrieve with slow figure-eight/hand-twist retrieve or a three-to-five-inch pull with a two-second pause between each repetition. These are the retrieves that best imitate swimming larvae, especially blue-winged olives and *Callibaetis*. These eddies can be found on most trout streams and might be the only game in town when water is at or below 40 degrees.

Time of day often correlates with success in cold-water conditions. It can also dictate the water targeted. On most of the water I fish in the winter, activity is squeezed into tight two- to four-hour periods that fall between 11:00 A.M. and 4:00 P.M. This is generally when water temperatures are at their warmest during the winter. I may be fishing prior to this period in an attempt to find "grabby" fish—those trout that are gluttonous or extremely aggressive. When I hit the afternoon sweet spot, I really start to concentrate on those riffle pools and seams that have depth sufficient for podding fish.

One other type of holding can come into play if water temperatures are warm enough (in the 39-to-42-degree range). These are flats that are

The Day-2 Midge Pupa is one of my favorite chironomid imitations when water temperatures are 38 degrees or lower. The most productive sizes are #16 to #22.

shallow in nature, between one and two feet, and have slow currents moving over them. Flats with these characteristics often warm faster than other parts of a stream. Aquatic invertebrates will become more active because of these warming water temperatures and become more available to trout. I will fish these flats with larval imitations of chironomids, blue-winged olives, and winter stoneflies like little black stones and little brown stoneflies. They are lightly weighted or unweighted because of the shallow nature of the holding water and the slow currents. This is also when I may turn to dry flies that imitate chironomids and blue-winged olives. I don't start fishing on the surface until I see a fair number of naturals on the surface or there are trout feeding on the surface.

As with riffles and seams, coverage of the water you are targeting is important. Temperatures might be slightly warmer on flats as air temperatures warm, but it is slight. Trout can still be sluggish because of slow metabolism. Several casts to holding trout can be required before a take occurs.

This brown trout was caught just downstream of a spring entering the main channel of a river. The spring was 4 degrees warmer than the river. Sometimes this small variation in temperature is all it takes.

As with warm-water conditions, trout often congregate at the mouth of surface and subsurface springs. In cold-water conditions, these fish hold at these confluence points because the water emitted is warmer than the rest of the stream. Springs are derived from subsurface aquifers and their temperatures are insulated from the cold air impacting exposed streams. Temperatures can be a few degrees warmer than the rest of the river. Sometimes these flow in at one point in the river. Others seep in at multiple points and are generally subsurface in nature. Finding these springs can lead to a gold mine of fish. When I am guiding on my home waters in the winter, there are specific springs I have access to that I will keep in my back pocket if the rest of the river is fishing slow.

Keep in mind that the term "spring" does not strictly mean cold water. Warm springs are also worth taking into consideration. On the lakes I guide on in Yellowstone National Park, particularly Yellowstone Lake and Lewis Lake, warm springs entering these stillwater bodies can be places of intense action during ice-out. Ice-out on some stillwater bodies produces excellent fishing, but there can also be inconsistency. I have had days on Montana's Hebgen Lake when ice-out has produced thirty fish in just a few hours and other days that were complete shutouts. Warm springs, however, can produce more consistency. The water near these springs can be as much as 50 degrees warmer than the rest of the lake. Lake water that is ten to twenty yards from the mouth of the spring can retain that perfect temperature for trout. This is where the action is. Generally there is still a significant amount of ice on the lake. The window of opportunity may only last for a few days before there is substantial ice melt and the rest of the lake's surface temperature warms to tolerable levels. Fishing then starts to pick up on flats, drop-offs, and lakeshores.

A QUESTION OF RELATIVITY

No two waters are the same. In past books I have consistently harped about how I change my fishing strategies or tactics based on stream gradient, lake composition and depth, forage composition, trout-species abundance, and a number of other factors. Water temperature is important in this regard as well. Wild trout on rivers like the Henry's Fork in

GET IT DOWN HERE!

When water temperatures fall outside their comfort level, trout descend deeper into the water column, a common survival mechanism. This can occur when water temperatures are *both* too warm and too cold. Holding further from the surface allows trout to find water temperatures that are more tolerable. It will be cooler when temperatures are hot. When temperatures are cold, deep portions of the water column can be slightly warmer. Furthermore, currents are slower at the deepest portion of the water column and at their fastest near the surface. Trout will avoid energy-depleting currents as when water temperatures are too warm or too cold.

These deeper portions of a stream or lake are where fly fishers should focus their attention when water temperatures are too warm or too cold. Most often, I am fishing nymph rigs to get to these parts of the water column. The issue, especially when fishing a stream as opposed to stillwater, is getting your nymph rigs down to where trout are holding before they drift overhead and out of the feeding lane, which can often be swift regardless of water temperatures.

This is where your tackle selection can play a big role. Heavily weighted nymphs seems like a no-brainer, but the fact of the matter is that when our nymphs hit the surface and then disappear as they descend, most of us think we are deep enough. The truth is, we have no idea where our nymphs are—they could be well above where trout are holding and feeding. When your nymphs are built with serious weight, you can be quite a bit more confident that the speed of their descent is sufficient. Nymphs with tungsten bead have been all the rage for well over a decade now—and for good reason. Tungsten is 1.7 times denser than traditional lead that was the standard a decade and a half ago. This does not necessarily mean that it is heavier, but what it does mean is that we can pack on one tungsten bead of a specific size for almost two lead beads of the same size. This is important when one considers pattern size.

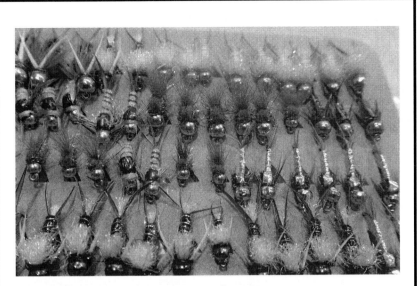

When fishing nymphs became part of my repertoire twenty years ago, I would use external weight like split shot to reach deeper portions of the water column. I now rely more on the weight of the nymphs themselves. Only when I need fast sink rates in deep water and swift currents will I turn to external weight.

For instance, when water temperatures are cold, I typically go with patterns that imitate chironomids, *Baetis,* and little winter stoneflies. Patterns I use to match these insects are typically between #16 and #22. It takes a lot to get these flies with so little mass down to sufficient levels in the water column. Tungsten beads can help out a lot.

Another form of tackle that fly fishers can employ to assist in sinking your nymph rig is a degreasing agent. I learned much about this through my discussions with former Fly Fishing Team USA members Pete Erikson and Jeff Currier. In the nymph-heavy international competitions like the World Championships, degreasing agents are applied to leader and tippet material so that they can break the surface tension faster and easier. A favorite of many fly fishers includes Loon Outdoors' Snake River Mud. This tool, along with heavily weighted nymphs, can get your rigging down where trout are holding and feeding.

Table 3.2: Water Temperature, Holding Water, Strategies, and Tactics

Water Temperature	Targeted Water
38 degrees and lower	**Streams:** Deeper portions of water column and current margins in riffle pools and seams, and confluence of springs and tributaries. Shallow flats at upper end of temperature range. **Stillwater:** Surface portion of the water column, flats, bank edges, and mouths of springs and tributaries.
39 degrees to 50 degrees	**Streams:** Most holding-water types, especially flats, riffles, seams, eddies, side channels, and confluences of springs and tributaries. **Stillwater:** Flats, mouths of tributaries and springs, bank edges, and submerged vegetation.
50 degrees to 63 degrees	**Streams:** All holding-water types, especially flats, riffles, seams, eddies, confluence points, tributaries, confluences, side channels, banks, and structure. **Stillwater:** All holding-water types, especially flats, bank edges, submerged vegetation, and drop-offs.
63 degrees and higher	**Streams:** All water types at lower end of temperature range. Deeper portions of water column in riffle pools, seams, eddies, and deep flats. **Stillwater:** All water types at upper end of temperature range. Deep drop-offs, submerged bars, and tributary mouths.

Streams: Dead-drift nymphs through deeper portions of riffle pools. Slow-retrieve nymphs at current margin of seams and at mouth of tributaries. Streamers retrieved slowly on floating and intermediate-sinking tips and lines. Concentration on afternoon hours from 12:00 P.M. to 4:00 P.M.

Stillwater: Slow retrieval of nymphs and baitfish imitations on floating, hover, and intermediate tips and lines. Concentration on afternoon hours from 12:00 P.M. to 4:00 P.M.

Streams: Nymphs and large attractors early in the day in favored water types. Imitative dry flies and nymphs during noon hours. Streamers fished with variable retrieval speeds on floating, intermediate, and three-inch-per-second sinking lines and tips in all favored water types.

Stillwater: Variable retrieval of nymphs and baitfish imitations on floating, hover, and intermediate-sinking lines and tips in all favored holding-water types. Dry flies when active feeding is observed.

Streams: Nymphs and dry flies dead drifted and/or swung through favored water types. Streamers fished with variable retrieves and retrieval speeds on floating, intermediate, three-to-six-inch-per-second sinking lines and tips.

Stillwater: Nymphs and baitfish imitations in all favored holding-water types. Variable retrieves and retrieval speeds on floating, hover, intermediate, and three-inch-per-second sinking lines and tips.

Streams: Deep portions of the water column in all favored holding-water types with dead-drifted nymphs. Large dry attractors with short dropper nymphs dead drifted or twitches in all holding-water types. Streamers with variable retrieves and retrieval speeds on intermediate and three-to-six-inch-per-second sinking lines or tips in favored holding-water types. Concentrating on early and mid-morning hours when stream temperatures are at their coolest.

Stillwater: Nymphs and baitfish imitations fished on hover, intermediate, and three-to-four-inch-per-second sinking lines or tips. Slow retrieval speeds or retrievals that mimic the forage you are imitating. Concentrating on early and mid-morning hours when lake temperatures are at their coolest.

Idaho or the Big Hole in Montana are conditioned to water temperatures through the evolution over generations that have come before them. I believe it is part of their DNA. What is too warm or too cold for these fish can be very different from what trout on New Mexico's Cimarron or Colorado's Roaring Fork tolerate. Similarly, trout on Yellowstone Lake might be used to temperatures that trout on Oregon's Crane Prairie Reservoir find intolerable. Don't make the mistake of taking the advice I give here and use it in a generic fashion for all waters. Thirty-nine degrees might be too frigid for trout on one stream, yet downright balmy for trout on another. Seventy-three degrees might mean near-death for trout on a certain lake, while trout on another might find such temperatures to be quite comfortable.

CHAPTER FOUR

NATURAL LIGHT CONDITIONS
Sunlight, Cloud Cover, and the Dark of Night

When we think of the picture-perfect scene of trout fishing any-where in the world, what comes to mind? Often it is similar to what we see on magazine covers, advertisements, and movies. There is warm sunshine, cloudless skies, and fly fishers decked out in nylon shirts and pants. It is a portrait of absolute comfort. What seems to make these days so perfect and comfortable is that glowing ball in the sky.

Nothing beats the comfort of fishing on those warm, sunny days that define trout water during the summer months. These idyllic days, however, do not guarantee good fishing.

These idyllic days are a reality. I have experienced them everywhere I have fished. A big drawing card on my home waters is that these are typical conditions during the heart of our fishing season from June through September. These kinds of days can be in the cards the rest of the year as well. I recall fishing the Snake River in Wyoming on a mid-April day in 1998 and having air temperatures peak at 77 degrees. Another memorable day was on the Green River in Utah in early March 2000 when the temperature was more than 65 degrees. Both of these days were accompanied by brilliant sunshine.

For every one of these perfect days, however, there are an equal number of days when the sun just isn't around. The skies we fish under are filled with clouds. There might be wind, precipitation, and cool temperatures. The glare from flat light on the surface makes visualizing your fly, be it a surface-riding pattern or a streamer below the surface, almost impossible. I have had these days on almost every stream or lake I have fished. I can remember boiling black clouds rolling over while I wade fished below Clark Canyon Reservoir on the Beaverhead in Montana. Just as memorable was the flat, gray sky that greeted me during two days of fishing Colorado's Frying Pan River one April. The sky seemed to be the same hue at 5:00 P.M. in the evening that it was at 7:00 A.M. earlier that day. If it wasn't for the rather good fishing I had on those two occasions, I would have chalked those days up as far less than idyllic.

Sunshine can make for a nice day on the water, but it does not guarantee good fishing. Conversely, clouds threatening on the horizon and then building throughout the day may not be the best situation one could ask for, but the fishing can be off-the-charts. Each of these conditions offer advantages and challenges. Each can offer solid productivity or downright awful fishing. Fly fishers can cope by employing a variety of strategies and tactics, some of which are based on the understanding of fish and the water they live in, and others based on tradition and lore.

NATURAL LIGHT CONDITIONS AND TROUT FORAGE

As with all natural phenomena, natural light conditions create their own situations that call for specific strategies and tactics. It can impact when

Caddis hatches tend to be synonymous with the warm, sunny days of May. However, I have had just as good surface action on those days when cloud cover dominates.

and where trout feed, as well as how and with what the angler fishes.

When we think about particular aquatic insects as they are emerging, specific types of weather generally come to mind. Take caddis for example. There are some 142-plus genera of caddisflies in North America. It is possible to see emergences of one species or another most months of the year. Despite this almost constant presence, the one caddis-oriented event that inspires fly fishers more than any other is the Mother's Day caddis hatch that occurs throughout the Rocky Mountain West on scores of streams each spring. On many waters, it is the unofficial start of the summer fly-fishing season. When you scan through the multitude of fly-fishing websites and print magazines that have covered this annual event, you will notice blizzard hatches occurring under bright, sunny skies. If you scan your own memory of fishing a caddis hatch in spring, most likely the same conditions exist as well.

The same can be said for most stonefly emergences. While we have winter stoneflies that can emerge under cold, cloudy skies in February and March, most species of stoneflies bring to mind warming-

Warm summer days spark images of damselflies and dragonflies hovering over the lakes we fish. Larval imitations can be fished no matter what the light conditions are.

weather trends. Skwalas are the quintessential spring emerger that comes out during that first, long span of warm, sunny days in April and early May. The big bruisers like salmonflies and golden stones, as well as their smaller cousins little yellow and little green stoneflies, are all about summer. On the streams where I guide like the South Fork of the Snake River, the Henry's Fork, and the Upper Green in Wyoming, these stoneflies emerge in a one-after-the-other succession from late May through July. Warm, sunny days are typical conditions when we fish these hatches.

Other invertebrates are synonymous with sunny skies. Damselflies and dragonflies are what I think of when I am fishing my favorite lakes in the middle of summer. I think the same about most terrestrials like grasshoppers, beetles, and ants.

But cloud cover brings many types of trout foods to mind as well, and perhaps none more so than a multitude of mayflies. Cold, cloudy weather in late autumn and early spring is when I see blue-winged olives emergences at their best. These conditions also spur the emergence of mahogany duns, a mayfly that means autumn has arrived on many trout streams. Pale morning duns emerge in a variety of natural light conditions, but it is under cloudy skies that I see the most intense hatches. And one of my favorite large mayflies, the great blue-winged red quill (most often referred to by its scientific name, *Hecuba*), is as tied to cloudy skies as it is to the water they live in.

Like almost all natural factors we deal with in fly fishing, there is nothing written in stone about light conditions and its influence on the availability of trout forage. Numerous personal experiences remind me of this fact. While I fish green drake emergences on the Henry's Fork in Idaho under sunny skies, one of the most intense hatches I have ever witnessed occurred just a short drive away on the South Fork of the Snake

River on a cool, cloudy day in early July. The Mother's Day caddis hatch that I typically link with sunny skies came to fulfillment for one mid-May day on Montana's Yellowstone River in 2007, but there was almost no surface activity to speak of until partly cloudy conditions took effect in the afternoon.

What really matters is not light conditions and the trout forage we *think* will be available. Rather, what really matters is the types of trout forage that are available on a particular lake or stream at a specific time of year and the conditions (generally related to water temperatures) that are associated with their emergence.

NATURAL LIGHT CONDITIONS AND TROUT VISION

Nowhere is the impact of natural light conditions more important than how it affects trout vision. Studying the ocular traits of trout can seem like an overkill topic for many fly fishers. In many ways it is. Anglers the world over ignore all but the basics of trout vision and do just fine. Nonetheless, understanding how trout see in various light conditions is a fun subject for inquisitive anglers to examine. It can also make you a better fly fisher.

In *Modern Trout Fishing*, I explored with a fair amount of detail the concept of light refraction and its influence on trout vision. This book does not call for the same in-depth coverage. However, there will be aspects from that examination that I will cover here.

To understand how natural light impacts the vision of trout, one must gain an understanding of the laws of refraction. A main dictate of refraction is that light

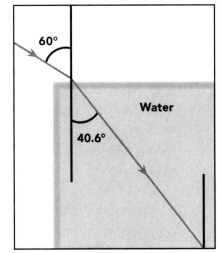

Light rays enter water between 10 degrees and 90 degrees. The tighter the angle to the surface, the more light will bend. An object that is 45 degrees or greater from a lake or stream surface may appear to be almost directly overhead to the fish below. (Image courtesy National High Magnetic Field Laboratory, Florida State University)

enters the surface of a stream or lake between 10 degrees and 90 degrees. When light enters the surface, it will bend. This bend is caused by the light moving from one medium (air) to another (water). Light travels through air much faster than it travels through liquid. This change in speed causes the bend in light. The abruptness of the bend is determined by the angle at which light is entering the surface. The tighter the angle to the surface, the more bend light achieves. Thus, light that enters at 30-degrees bends more than light that enters at 45 degrees but bends less than light that enters at 20 degrees. Light that enters at 90 degrees (straight down) will not bend at all.

What is interesting about refraction is how objects above the surface appear to trout based on the angle that light is entering the surface. If an object is at a position where its image is entering the surface at 10 degrees, the bend caused by refraction makes the object look much higher than it actually is. The image of objects that are entering the surface at 45 degrees or greater appear almost overhead to a trout. We can only assume that trout would be spooked out of their skins if an object is right on top of them.

Two factors can limit what trout see above the surface. The first of these is surface disturbances caused by wind or currents. This can cause breaks in the surface that disrupt the visuals trout can get on above-surface objects. Types of holding water where this occurs include riffles, seams, eddies, confluence points, and submerged structure. Some are more disturbed than others based on gradient and stream volume. This is where the fly fisher has an advantage. Think about how close you can get to trout in these holding-water types as opposed to shallow flats on a stream or possibly any kind of holding water on stillwater. The few times when there is less need for stealth in these holding-water types is when wind causes disturbance.

The other factor that limits what trout can see above the surface is lack of natural light. There is substantial evidence that the vision of trout has limitations once things go dark. Quite simply, the more light there is, the better trout can see. One particular study suggests that the reaction distances of coastal cutthroat trout increase rapidly with increasing light intensity.

Brown trout—the predator at dawn, dusk, and all through the night.

This is not to say that trout have the same difficulty seeing at night as do many terrestrial creatures, including humans. There is substantial evidence of their keen ability to intercept forage and avoid predators in low light conditions. Part of this is based on their use of highly tuned auditory and olfactory mechanisms, but much of this is due to an ocular adaptation evolved to ensure survival in low light.

Differences exist between trout species in their feeding abilities during low light conditions. Perhaps none is better at feeding in the dark than the brown trout. Studies have shown that brown trout (along with brook trout) had maximum scotopic (vision of the eye under low light conditions) thresholds two times lower than rainbow trout and cutthroat trout, meaning that they are more active at night, twilight, and under conditions such as heavy cloud cover. Rainbow and cutthroats will feed at night, but typically when there is light provided by celestial bodies like the moon and starlight. Other studies suggest that brown trout favor, and are more active in, shaded areas of streams than other species. These studies reinforce the image of brown trout as both the ultimate nocturnal predator as well as the big toad always hiding behind the log, ready to ambush its prey.

While we may observe more feeding by rainbow and cutthroat trout in sunlight than brown trout and brook trout and, conversely, more activity by brown trout and brook trout in low light conditions, the simple matter is that all these species will feed in both conditions. I was raised in cutthroat country and still reside there today. I have caught plenty at and after dusk. The same can be said for lake trout. I live very close to two exceptional lake trout fisheries: Jackson Lake in Grand Teton National Park and Lewis Lake in Yellowstone National Park. These fish have the reputation for being deep dwellers that reside at depths that are out of reach for the fly fisher. This is not the case on the lakes I fish. I guide fly fishers on these lakes a few times a year. When twilight comes and the darkness of night approaches, the action at or near the surface can be exceptional.

There are variations in where trout hold based on light conditions. On Wyoming's Snake River, trout typically hold tight to current breaks like riffles, banks, structure, and the extreme upper portions of seams and confluence points. The Snake is a high-gradient stream, so holding tight makes a lot of sense. But on those days when there is significant cloud cover, I notice trout holding a noticeable distance from their

Trout can hold well off of prime lies when cloud cover prevails. Fish like this Snake River cutthroat may feel more protected from predators with less sunlight.

favored lies. Instead of holding right at the upstream edge of a riffle, they can often be found several meters downstream in the slower, shallower riffle pool. In this kind of water, trout are exposed to predators like ospreys, eagles, and pelicans. But when there is low light, trout will be more difficult to see below the surface. I believe that trout hold here for that reason. They feel a bit more protected from these airborne predators. I attack these more exposed waters with streamers and moderately sized attractors when the skies have significant cloud cover.

I have also observed differences in feeding activity when there is cloud cover. One particular day on the Madison below Hebgen Lake stands out. During a mid-May caddis hatch, clouds were passing over the water intermittently throughout the day. It was the purest definition of "partly cloudy." When the cloud cover shielded the stream, rainbows were popping the surface with intensity. However, when the sun came out for a fifteen- to twenty-minute period, surface feeding would subside. This occurred throughout the day. There was no change in the number of bugs on the water, but the trout showed no interest in them. The change I experienced in feeding activity on that day could be explained by the instinctual reaction trout have to being more visible to predators.

Size and color of your pattern is an important consideration when low light conditions are a factor. The size of the pattern seems almost trite, but using large patterns can increase productivity when fishing after nightfall. The effectiveness of large patterns after sundown is the consequence of a number of elements working together. The first is that trout, particularly brown trout and brook trout, lose much of their apprehension in low light conditions and become more active. The research highlighted in this chapter suggests this.

Large patterns are also easier to see in darkness. Think about how productive stripping a size 2 baitfish imitation through a lake or stream or skittering a mouse pattern across the surface of a stream can be under night skies lit only by the moon or stars—or with almost no light at all. Such large patterns do not often show up under clear skies because they are too big. Some anglers claim there is too much for trout to look at. Imperfections—most notably the large piece of metal sticking out its abdomen—are easily detectable. But in darkness or low light conditions,

Mice and lemming patterns are effective at night. They are easy for trout to see, to cause an easily detectable surface commotion on the retrieve, and to imitate something that supplies a lot of calories.

these imperfections are more difficult for trout to see. What is more, these large patterns typically make a commotion. A big streamer stripped through stillwater will make noise that trout easily notice. The same can be said for mouse patterns as they are stripped across the surface of a stream.

Many types of trout forage that are available at night happen to be large specimens. Rodents, for example, are nocturnal by nature. When mice or lemmings happen upon a body of water, it is usually at or after sundown. Many larger aquatic insects emerge at night or return to the stream surface at night to deposit eggs. Perhaps none is as well-known for this as the giant, inch-and-half-long *Hexagenia*. There are dozens of famous photos from the Upper Midwest featuring thousands of *Hexagenia* congregating at the bright headlights of streamside vehicles.

Pattern color is something I consider less important than others aspects of trout food imitation. In fact, in *Dave Whitlock's Guide to Aquatic Trout Foods*, the author states his belief that fly color is less important than size, silhouette, movement, and event texture. But what Whitlock is describing is the importance of these factors to *imitation*. What I explain here is pattern color in regard to how trout react to it under low light conditions.

As with human eyes, a trout's retina is composed of a set of rods and cones. Cones are used for bright light conditions and allow the eye to see images in color. As low light conditions increase, the cones retract and light-sensitive rods extend. It is not just a cloud-filled or completely dark sky that causes this reaction. Depth in the water column can cause this as well, because ultraviolet-light penetration diminishes with depth. Discolored water can do the same. These rods allow trout to detect objects in black-and-white. Thus, as light diminishes, the ability of trout to detect the color of objects lessens. Most of the scientific literature on this matter suggests that, no matter what the natural color is, objects will appear in either black or white.

There is a rule of thumb in angling circles that says "Bright day, bright fly. Dark day, dark fly." Our understanding of trout vision lends credence to this adage. Colors on the red side of the color spectrum—red, orange, and yellow—appear brilliant and vivid to trout under conditions of intense light. The brilliance of these colors is reduced with distance and diminishing light. The less light there is, due either to decreasing natural light, depth, or discolored water, the more that objects composed of these colors appear black in color.

At first glance, it seems this would suggest that any color or pattern would work in low light conditions because it is going to appear dark to the trout. If, however, we take into account colors other than red, orange, and yellow in the color spectrum, and how trout see contrasts in these colors, we see that color can still be important when natural light is low.

An adage states "Bright day = bright fly. Dark day = dark fly." I do not blindly follow this rule of thumb, but I find that dark streamers tend to perform best on days with heavy cloud cover while bright streamers produce better on bright, sunny days. My streamer box is stocked for both occasions.

In low light conditions, streamers that are dark in color tend to outperform brighter patterns. Dark baitfish imitations like Kelly Galloup's Boogeyman or the Lite Brite Zonker create a dark shadow against a dark sky to fish looking up. The size of a dark streamer comes into play based on other factors and stream conditions. Note that both patterns here are tied with a Rapala loop, which gives them independent movement on the retrieve.

Colors on the violet side of the color spectrum, particularly blue and purple, do not appear near as brilliant to trout as colors on the red side of the spectrum. Yet blue and purple have three advantages over other colors for the fly fisher. First is the fact that these colors can be detected at far greater distances than other colors. According to research, colors such as red or orange can appear as almost an opaque blur at certain distances. It is not until an object containing these colors are within several feet of a trout that its brilliance comes through and the true shape of an object can be distinguished. Blue and purple objects, on the other hand, can have a more detectable shape from many more feet away.

Another advantage to colors on the violet side of the spectrum is that trout have the ability to recognize differences in hues. Trout can see shades of indigo, plum, and purple in an object better than they can see shades of scarlet, coral, and cherry.

But more important is the fact that colors such as blue and purple are the last colors trout see as light diminishes and the first color they see when light returns. What this means is that at twilight, objects on the red side of the spectrum may appear ill-defined, black, or maybe even invisible, while objects that are blue or purple appear much more defined.

I am a fan of matching the color of the natural forage I am imitating. If I am fishing a mahogany dun adult through a riffle in late September, most likely it is going to have a coffee-colored abdomen and a smoky wing. My caddis larva in late spring will have contrasting hues of green or cream, depending on the particular specimen I am matching. The streamer I am fishing throughout the year will typically have the dark dorsal side, the light ventral side, and the lateral line found on almost all forage fish that make up a trout's diet.

But natural light conditions often do come into play. When they do, I will not shy away from colors that are far less imitative. In those midsummer months when the sun is full and bright, attractors like a winged Chernobyl, Double Humpy, or Stimulator with red, orange, or yellow bellies can be productive searching patterns when fished along banks and structure, and even in a riffle or a seam. This is true on many of the waters where I fish and guide. I have just had too much luck over the years with these colors.

And when things are dark due to heavy cloud cover or the low angle of the sun at dusk and dawn, those colors on the violet side of the spectrum almost always become important parts of my arsenal. A dry fly like Carlson's Purple Haze is one of my favorite patterns when the sun dips below the horizon. The Batman, an unorthodox blue-purple-black nymph originally distributed through Idylwilde Flies, is a pattern I turn to on those days when the sky is full of dark clouds and there is no surface action. During those special moments in the evening when there are more stars in the sky than there is sunlight, a jet-black Quad Bunny becomes my streamer of choice. I will fish it until I can't see the tip of my rod.

IT'S NOT JUST ABOUT WHAT THE FISH SEE: NATURAL LIGHT CONDITIONS AND THE FLY FISHER

Natural light conditions can influence the angler just as much as they can influence the fish. Yes, we have the aid of polarized eyewear that allows us to see the flies and rigging we are using as well as the trout we are targeting. Nonetheless, there are those times when our eyes fail us, even with the help of $250 sunglasses. The sun is too bright, the light is too flat, or it's just too dark.

There are certain tricks fly fishers deploy to deal with these situations. Guides in particular have commonsense tools to make it a bit easier for their guests to see what is going on. One of the simplest is to position the angler in a way that the glare of natural light is significantly limited. This could be downstream of the targeted

Light stops penetrating water at 10 degrees. I find getting low to the water (below the 10-degree angle from the angler's eyes to the fly or indicator) greatly reduces the glare that impedes a fly-fisher's vision.

holding water, or perhaps along the flank of the holding water and in a way that the directional flow of natural light is at the back of the person casting.

Sometimes, however, gaining such a position is not possible. The water may be too deep or the casting lane obstructed. This is where getting low to the water can assist a fly fisher in seeing the rigging. When I am guiding a guest and glare or flat light is just too much, I will drop to a knee, with the surface current flowing at diaphragm level to better gauge the drift of the rigging. Remember that light stops penetrating the surface at 10 degrees. When I get at a low angle in order to see my rigging, my vision is typically below this 10-degree threshold.

Nothing does more to assist the fly fisher in seeing his or her rigging than the composition of the rigging itself. Wing material on a dry fly is a perfect example. Most wings are composed of lightly colored material to assist anglers in detecting their fly. Common materials include calf tail, EP Fibers, and Poly Yarn. White is the most common color, with pink, yellow, chartreuse, and red also popular. But there are times when the glare of the sun or flat light on a riffle renders these colors ineffective. This is where darker colors can come to the rescue. The use of black calf tail on a Parachute Adams, referred to as a "Glarachute Adams," is a remedy for cutting through glare that might have been around since before I started guiding in the early 1990s. The basic concept is that dark colors like black can cut through glare in a way that lighter colors cannot. For some patterns, going with a darker wing can be more imitative than a lighter-colored wing. This is the case with mahogany dun and brown drake imitations and, to a lesser degree, *Callibaetis* and March brown patterns.

The idea of the Glarachute is not reserved for surface patterns alone. It has a role in nymph fishing as well. In the Rocky Mountain West where I guide, nymphing is done primarily with a floating or near floating indicator. These devices assist in detecting strikes and suspending nymph rigs. Indicators come in many forms.

Table 4.1: Natural Light, Holding Water, Strategies, and Tactics

Natural Light Conditions	Targeted Water
Sunlight: Sunny, mostly sunny, and partly cloudy conditions	All holding-water types on streams and stillwater, favoring deeper lies when surface activity is negligible.
Cloud Cover: Overcast, cloudy, and mostly cloudy conditions	All holding-water types on streams and stillwater. Water column near the surface can show increasing signs of activity. Target slower, shallower portions of a stream or lake that is a good distance from typical lies that are deeper and have swifter currents.
Dawn and Twilight	All holding- water types on streams and stillwater. Potential for decent activity at or near the surface in riffles, seams, eddies, confluence points, and flats on streams. Drop-offs, flats, bank edges, submerged vegetation, and tributary outlets on stillwater
Predawn and Post-Twilight	All holding-water types on streams and stillwater. Potential for decent activity at or near the surface in riffles, seams, eddies, confluence points, and flats on streams; drop-offs, flats, bank edges, submerged vegetation, and tributary outlets on stillwater.

Dry flies and nymph patterns:

Stonefly, caddis, dragonfly, damselfly, terrestrials, and some mayfly imitations. Colors on the red side of the color spectrum: red, orange, yellow.

Streamers:

Baitfish imitations in bright hues: gold, yellow, copper, chartreuse, etc. Retrieve with speed and erratic movement.

Dry flies and nymph patterns:

Stonefly, chironomid, dragonfly, damselfly, and terrestrials imitations, but trending toward mayfly and caddis imitations. Colors on the red side of the color spectrum but trending toward colors on the violet side of the spectrum: blue and purple, as well as naturally dark colors.

Streamers:

Baitfish imitations in bright hues, but trending toward darker hues on the violet side of the color spectrum.

Dry flies and nymphs patterns:

Caddis and mayfly imitations primarily. Some activity on stonefly, chironomid, dragonfly, damselfly, and terrestrial imitations. Colors on the violet side of the spectrum: blue and purple, but also black.

Streamers:

Baitfish imitations with darker hues on the violet side of the color spectrum: blue, purple, dark gray, but also black.

Dry flies and nymphs patterns:

Large patterns imitating stonefly, caddis, dragonfly, damselfly, terrestrials, and mayfly. Colors on the violet side of the spectrum: blue and purple, but also black.

Streamers:

Baitfish imitations with darker hues on the violet side of the color spectrum: blue, purple, dark gray, but also black.

Some are plastic balls called Thingamabobbers, some are yarn puffballs, and some are composed of colored leader material referred to as sighters. Generally these indicators are brightly colored in red, pink, orange, yellow, or white. But just like light-colored wing material on dry flies, these colors can become lost in glare. To compensate, many fly fishers over the past several years have turned to darker, sometimes asphalt black, indicators. Just like the Glarachute, these darker indicators cut through sunshine and flat light glare better than the more traditional colors. My favorite indicator is a black Thingamabobber between ¼ inch and 1¼ inch in diameter, depending on the situation. I actually use a black Thingamabbber in every natural light condition when I am nymphing with a suspension device.

Surface patterns with dark parachutes or wings are often referred to as "Glarachutes" because they can be easier to see in flat light and bright sunshine. Those fishing nymphs can use black indicators when light conditions are difficult.

CHAPTER FIVE

WATER LEVELS
AND CLARITY

In 1993, I was in my third year of guiding and a certified know-it-all at age twenty-two. I often worked solo with single-party trips going where I wanted, when I wanted. Taking a note from earlier experiences, I avoided the high, off-color water of early season runoff. If a stream wasn't clear or close to clear you would not see me on that water. In my mind, trout needed clarity to eat flies.

Fishing high, off-color water can be demoralizing, but trout can still be caught. Focusing on the right water and the right patterns will turn a potential shutout into a good day on the river.

So it is easy to understand my disgust when I was forced to join my father on a two-boat trip down the Green River in Sublette County, Wyoming. It was early June, and the Green was running the color of cardboard and 40 percent above normal flows.

What is my father thinking? I would ask myself. *Why is he doing this to me?*

My feelings were the farthest thing from my father's mind. We were there for a reason. He had fished the Green in these conditions before and, while there was no guarantee on this given day, he knew how good it could be. And it certainly was good. Despite high water levels and off-color water, our parties were able to hook into dozens of brown and rainbow trout with large surface attractors and streamers, each fished with a lot of movement. It's hard for me to comprehend my ignorance, let alone the arrogance, of those early years of guiding. But it sometimes takes a bit of force by the guy signing your checks to get you out of your confidence zone. You discover that you don't know everything there is to know about fishing, even if the stream in question is almost in your backyard.

Trout fishing tends to be synonymous with crystal-clear water. It is the image of purity that we link trout to in the first place. But I have fished and guided on enough streams with high flows and off-color water to know that, with a slight change in tactics, the fishing can be very good. Yes, there are limits to production in high and off-color water. One of my home rivers—the Snake in Wyoming—usually fishes best at between 1,800 and 6,500 cubic feet/second (cfs). Yet I have seen it peak at a record 38,000 cfs in 1997. The adage "too thick to drink but too thin to plow" was fitting. You would be hard-pressed to catch anything in those conditions.

I am familiar with several steelhead streams with the reputation of fishing best at high flow levels, off-color, or both. I fish for carp and bass on some streams and lakes that are in their prime when the water is murky. Most fly fishers who go after these game fish understand and accept these facts. Yet their hopes and aspirations seem to plummet when the stream they are targeting isn't in pristine condition. Trout can still be caught in high water. And sometimes the fishing can be epic. What the fly fisher needs to do is go in with confidence and be prepared to change tactics if tried-and-true techniques are failing.

Big attractors can be the ticket when streams are running at higher than average levels. They can be suggestive of natural insects and they are easier for trout to see. A dropper nymph coming off of a big attractor is another good way to go. This big brown trout was caught on the upper Madison with a #8 foam attractor when the river was flooding the fringe of its vegetated banks. It helped that the reach we were on was experiencing a steady hatch of salmon flies.

FINDING TROUT IN HIGH WATER

A stream with a high water level will typically hold more trout than a stream with low water levels. If this is the case, then why is it that fishing seems so tough when flows are high? One reason is that there is simply more habitat for trout when water levels are higher. I will use a favorite riffle of mine on the South Fork of the Snake River in Idaho as an example. When flows from Palisades Reservoir are running at 6,000 cfs, approximately 30 percent of the cobblestone bar forming this riffle is exposed. Of the remaining 70 percent, about half of it has less than one foot of water flowing over it. In conditions like this, trout are concentrated in the small riffle pool located on the downstream margin of the bar. The riffle pool is what I am targeting.

Higher-than-average flows allow trout more water where they can hold. This fly fisher hooked into several rainbows and cutthroat feeding on a cobblestone shelf at the head of a long, deep riffle.

Now consider flows that are more than twice this level—13,000 cfs. In these conditions, the entire cobblestone bar is submerged. Some of that bar has as much as two and one half feet of water flowing over it. The result is a significantly larger riffle pool downstream of the bar. Trout holding in this pool have a larger area to hold and feed. The tail-out of this riffle extends much farther downstream, allowing more trout to hold in this portion of the pool. Furthermore, the bar forming the riffle creates additional holding water. The South Fork of the Snake River has a moderate gradient. The riffle I am examining exists where the stream gradient is approximately nine feet per mile. Such a gentle slope to the streambed allows trout to hold *on* the bar when there is sufficient depth, and at 13,000 cfs there is more than enough depth. Trout holding on the bar can get the first crack at drifting invertebrates and emerging insects.

These higher flows do not render a riffle like this unfishable. It can be fished, and sometimes the fishing is very good. There are some excellent fly fishers who prefer fishing the South Fork of the Snake River at these higher flows. But you have to be more focused and fish with more diligence when flows are higher. Add off-color water to the mix—a factor often associated with high water levels—and your focus must be taken to another level.

The impact of off-color water on fishing is obvious: potential foods for trout are more difficult to see. Trout have exceptional vision and auditory systems, but they still struggle to identify forage when visibility is reduced. Your #10 Winged Chernobyl or #16 Hares Ear Nymph might be in what you think is the right place, but that doesn't mean that trout can see it.

The first task to successfully fishing high and off-color water is to find the places where fish are holding and feeding. This seems simple enough. In fact, this might be the simplest part. Riffles, seams, eddies, confluence points, and banks will all hold trout regardless of flows or clarity. But in high water they are more dispersed. Thoroughly covering

This Madison River rainbow was caught while the stream was high and slightly off-color. We fished the slower current margins that day and had quite a bit of success.

these places where trout hold is crucial. This can be said about fishing in ideal conditions, but it becomes much more important when flows are high and water is off-color.

There are places in a typical piece of holding water that should be primary targets. The first of these is related to the water column. When water levels are high, there is often higher stream volume and faster currents. We know from rules of hydrodynamics that currents are faster near the surface than they are near the bottom of the water column. When currents on a stream are fast, trout are typically not holding in the same place within a riffle or a seam as they would under normal conditions. Trout can be lower in the water column because currents are slower. While it is good to target deeper portions of the water column, don't make the mistake of going too deep. Trout are generally looking up to find their prey. I have heard the same from other trout fishers like George Daniel of the US Fly Fishing Team and steelhead anglers, such as Dec Hogan. If you are fishing nymphs or streamers and your rigging is too deep, your flies might be below the fish you are targeting, and they might never see it. If your flies are slightly elevated in the water column, trout will be able to see them much easier. It is far easier for trout to see potential forage against a bright sky at the surface than against the dark substrate of a stream bottom.

The next consideration is where trout are located horizontally along a piece of holding water when streams are running high. Finding slower currents is the goal. Even under conditions with high and off-color water, you can still find slow currents in the proximity of prime target water. I will use the riffle on the South Fork I discussed earlier in this chapter as an example. The current flowing over the bar at the head of this riffle is faster than it is a few meters downstream in the riffle pool, and the current at this portion of the riffle pool is faster than it is at the tail-out. Trout will be holding all along the riffle, including on the bar at the head of the riffle, but the tail-out is where I generally find trout in great concentration. It is the tail-out that I focus on. This goes for other types of holding water with tail-outs, such as seams and confluences. You can also find slower currents along banks. Not only can currents along banks be slower, the water along the bank margin can be noticeably clearer. Trout holding along the bank can see your flies easier than they can farther out in the stream.

Water that lacks distinct tail-outs is also worth targeting. A good example would be a rock garden along a bank where there is lots of submerged structure. In this case, trout will not be dispersed like they are on riffles and seams in high water. Instead, they will be concentrated at the slower spots directly downstream of boulders and large rocks. That slow water can be easily detected by the current break being created on the surface by the submerged structure. Sometimes the current break is only two square feet in size, but that is large enough to hold sizeable trout. This water is an important target when water levels are high and currents are fast.

I have always said that the water you fish is more significant that the pattern you are fishing with. This maxim holds true for me no matter what conditions I

Submerged structure, be it woody debris or rock gardens, provides fun pocket water to fish no matter what the conditions happen to be. During periods of high water levels, these features can be crucial. The small amount of current breaks they provide can support a large number of fish.

am fishing in. When fishing high and off-color water, however, the pattern I am fishing (and the way I fish it) becomes more important.

Larger patterns are easier for trout to see in high and off-color water. A big attractor skittered across the surface of a trout stream generally gets the attention of trout far more often than a #16 caddis imitation. This is the case below the surface as well. Large streamers and stonefly larva patterns attract trout better than smaller mayfly and caddis flies. I may fish a stonefly nymph like a Pat's Rubber Leg in tandem with a smaller pattern like a Lightning Bug, but unlike fishing in more ideal conditions when the

Dark streamers tend to create a shadow effect in off-color water, which many anglers think makes them easier for trout to see. Some fly fishers also believe that bright streamers attract trout in clear water because of the play of radiant light on vividly colored materials.

smaller pattern takes two or three fish for every one taken by the larger pattern, it is the bigger nymph that often racks up the numbers.

Color, silhouette, and movement of your patterns are just as important as size when fishing high and off-color conditions. In chapter 4, I discussed the old adage "Bright sky—bright fly. Dark sky—dark fly." Another old saying is "Bright water—bright fly. Dark water—dark fly." Too much emphasis can be put into these dictums. Yet there are reasons why so many fly fishers put their faith in them. One argument is that many kinds of prey attempt to blend in with their surroundings. Perhaps the most valid reason for me has to do with the fact that dark patterns create that shadow effect against a dark background (in this case, the background being the off-color water). This allows the silhouette of a dark fly to be much more visible than that of a brighter-colored fly. Test out this theory for yourself. When your local stream goes muddy, find a pool and compare the visibility of a black sculpin pattern to that of a brighter streamer, such as a Tequilly or J.J. Special. You should notice that the darker fly is seen at a greater distance than the brighter one. Research on the ocular traits of fish has shown the same for trout. Dark colors like purple and, to a lesser degree, blue are ones that fish can see at greater distances than other colors in low light conditions. These colors are the darker ones on the color spectrum, so they, along with black, will stand out with that shadow effect against the dark background of muddy water.

The "dark water—dark fly" mantra is not exclusive to trout fishing. I know of a number of steelhead anglers who use darker patterns when the rivers they fish are muddy. The same goes for some saltwater anglers as well. My friend Will Dornan fishes out of Islamorada, Florida, every May and June. He uses dark tarpon flies when fishing brackish estuaries and bocas. Renowned bonefish stalker Dick Brown suggests using patterns that blend in with the bottom one is fishing—light-colored flies on bright, sandy flats and dark-colored flies on marl and flats with vegetation.

I experience a lot of production with dark patterns in high and off-color stream conditions. Nonetheless, bright-colored flies can work just as well. Colors like red or pink cannot be seen at great distances or in low light conditions, but as they enter a trout's field of vision, they are vibrant. Thus these colors are in play in high, off-color water conditions. This is why red or pink San Juan Worms or a red #10 Copper John can be so effective when streams are high and off-color.

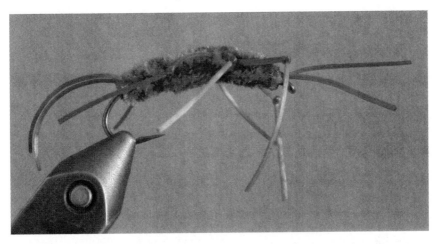

Despite the "dark water—dark fly" rule of thumb, bright-colored patterns in red, pink, or orange are effective. These colors cannot be seen from long distances, but when they enter a trout's field of vision, they are vibrant. Patterns like an orange Pat's Rubber Leg (pictured here) have the added quality of animated material movement, which can attract the attention of fish when streams are high and off-color.

Pattern movement often generates responses from trout when visibility is low. Rapid retrievals and erratic jigging of streamers in muddy water can produce good results. The same goes for nymphs that are jigged or swung through holding water.

The movement of materials on dead-drifted nymphs can also generate strikes. Consider the movement of chenille on a San Juan worm, the Flexi-Floss legs on a Pat's Rubber Leg Stonefly nymph, or the hackle on a large soft hackle caddis pattern. Stream currents impart movement in these materials when they are dead-drifted. This movement is easily detected by trout as they drift near their lie. That movement alone can be enough to produce strikes.

FINDING TROUT IN LOW WATER

Low water levels on streams and sometimes lakes are seen as ideal conditions. Water is clear, which means trout can see your flies easily. Furthermore, fish are concentrated in pockets of holding water. They will be podded up in riffles, seams, and flats.

Successfully targeting spooky trout does not require long-distance casts. A deliberate approach and picture-perfect presentation are much more important. This Owyhee River brown trout was caught in less than a foot and a half of water and a brutally slow current. The angler stood less than twenty-five feet away when the fish was hooked.

But shallow water produces its own challenges. In low, clear water, fish are far more vulnerable to predators. They can be on high alert and spook easily. A fouled cast or a less-than-desirable presentation is all it takes to put trout down.

For me, fishing in low-water conditions is all about presentation. Getting into position and approaching your target slowly is key. I start with short casts and work line out in three-to-five-foot increments. Once I have let out between thirty and forty feet of line, I begin to slowly work my way in the direction I am casting. No more line is coming out of my reel. All further progression toward my target is performed with slow, methodic footsteps in three-to-five-foot increments. I use this tactic because a line that is short or moderate in length is easier to control during the cast. It is less likely to foul and the angler can better control acceleration and power of the cast. A shorter line is also less likely to be disturbed by cross currents that can hinder your desired presentation.

Some may think that a thirty-to-forty-foot cast is too short for spooky trout in shallow water. But remember that the angler has a bit of an advantage. Shallow water means that trout are closer to the surface. The closer trout are to the surface, the less they can see through their cone of vision. This is something I discuss in my 2013 book *Modern Trout Fishing*. The only time I will work out more line is when I am sight casting to a specific trout, and the surface currents guarantee the drift or movement I desire.

When I am working upstream, I cast only my leader over the fish or the water I am targeting. The line stays out of their cone of vision. At the same time, I am still casting far enough upstream so I can get the proper drift over the target. I do not apply any kind of formula in terms of distance of the line-leader joint to the target. What I generally find is that if a trout is holding in two feet of water, the line-leader joint can land on the surface approximately three to four feet from my target without putting the fish down. If trout are holding another foot lower, I increase my distance by another two feet. If the surface current is broken by subsurface structure, a riffle, or a seam, the line-leader joint, as well as the angler, can get closer to the trout because its vision will be disturbed.

I use the same tactics of approach when fishing downstream that I use when fishing upstream: three-to-five-foot increments of line with

When working downstream during periods of upstream wind, a draw-back cast can be employed, allowing for precise presentation without fighting through a stiff breeze. Cast just upstream of your target water with slack and allow your rigging to drift downstream to the trout. After the drift is complete, simply draw your line back upstream and allow it to drift back downstream again by using the slack and slowly lowering your arm in a controlled fashion

each cast, work out between thirty and forty feet total, and then start working downstream with slow, methodical footsteps in three-to-five-foot increments. What changes is the type of cast I am using. When working downstream, I perform a draw-back cast (also referred to as a parachute cast or pile cast in some angling circles). With the draw-back cast, I do a typical overhead cast with the full length of line. Once the line reaches full extension on the forward stroke, I bring the rod back upstream before the line and leader make contact with the surface. Essentially, I am drawing the cast back upstream before it lands on the surface. The result is a line with enough slack on the surface to allow the rigging to drift down to the target without dragging or creating an undesirable wake.

FINDING TROUT IN RISING WATER LEVELS

Like many fly fishers who target trout, I consider increasing water levels to be among the most difficult conditions one can fish. Ascending water levels can be caused by snowmelt during spring runoff, excessive precipitation, or increased flows from reservoirs. Each of these can lead to off-color water created by streamside sediment being picked up and carried downstream as water levels increase. But this is only part of the challenge. With rising levels, stream volume is also increased. This allows trout to disperse throughout more portions of holding water. Riffles, seams, and eddies are bigger. And more than anything else, increasing water levels are a major change to a trout's environment. It takes time for fish to acclimate to the changes they are experiencing.

Despite the odds anglers face when water levels are on the rise, there are certain things that play to our advantage, and most revolve around the kinds of forage available to trout under such conditions.

When rivers rise due to increased releases from dams, what is in the reservoir will end up flowing out if it. This includes high-protein crustaceans trout crave, like scud and mysis. These food forms are always entering tailwaters if dam gates are open and flowing, but their entry is intensified when flows increase. Reservoirs throughout the United States contain these creatures. Many state wildlife management agencies stocked mysis and scuds on their lakes specifically as forage for stocked game fish. My home tailwater stream, the South Fork of the Snake River in Idaho, is a good example. Flows from Palisades Reservoir, which feeds the South Fork, generally peak in late May and early June. The river can be discolored and dry fly fishing is tough, but the nymphing action can be solid. We throw large stonefly larva patterns during this time of the year because our

All-purpose nymphs like the Lightning Bug are used as mayfly larva imitations. Some fly fishers also use it as a scud or mysis shrimp imitation on tailwaters like the Frying Pan in Colorado and Montana's Bighorn River.

salmonflies are becoming active. But trailing behind this bigger fly is something that is suggestive of a mysis or scud. Popular patterns include Landon Mayer's Mysis or a Dunnigan's Scud. Some anglers suggest that all-purpose nymphs like the Lightning Bug or a Rainbow Warrior are imitative of scuds and mysis. All of these patterns in #10 down to #20 are used on reservoir-fed rivers throughout the Rocky Mountain West like the Frying Pan in Colorado and the Bighorn in Montana.

Rising water levels can also lead to more terrestrial-oriented forage on the water. Be it increased releases from a dam or heavy rains flooding a riverbed, rising water washes over normally dry banks that are populated by worms, ants, beetles, and all kinds of other land-dwelling creatures. The faster the water rises, the more of these trout foods can end up in a stream. While trout are acclimating to increased water levels, they are still opportunistic animals. You do not have to fish a specific pattern. Instead, focus on fishing those places where these creatures are found. Banks and exposed structure are good choices.

As rivers recede, aquatic invertebrates like this caddis larva are forced to move from drying parts of the streambed. This migration renders them more vulnerable to feeding trout. This is why lightly weighted nymph rigs can be a good way to go in such conditions.

FINDING TROUT IN DESCENDING WATER LEVELS

As with rising water, trout must acclimate to stream flows when they are decreasing. There is variability between streams. I have experienced excellent fishing on Wyoming's Snake and Green Rivers when water levels are going down. I cannot say the same for the Box Canyon reach of the Henry's Fork below Island Park Reservoir in Idaho. When flows are decreasing on this piece of water, fishing can be tough. Differences in action many times come down to the rate of recession a stream is undergoing. A slower pace often leads to eas-

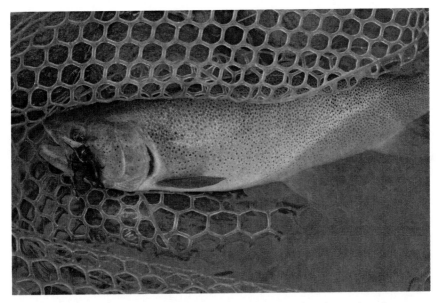

Forage fish can become displaced when streams recede. These chaotic events force them into the lies of dominant trout. Baitfish can become easy prey. This is a big reason why streamers can be deadly when rivers are on the drop.

ier acclimation by fish, while faster rates can lead to an uncomfortable change to a trout's environment.

Nonetheless, the general consensus among fly fishers is that fishing is better when flows are descending, not when flows are going up. In my opinion, the reason fishing is good in descending water is due to the changes trout and their forage experience under such conditions.

As water levels recede, aquatic insect larva must retreat from those parts of the streambed—side channels, shallow riffles, flats, and troughs along banks—that are susceptible to going dry. This is the case with all types of waterborne invertebrates, not just stoneflies, mayflies, caddis, and chironomids, but also crustaceans and leeches. As they retreat, these creatures become available to trout. Excellent nymph fishing is often the result. I target the mouths of side channels, banks, the drop-off edges of shallow flats, and the heads of riffles under these conditions. I often use a double nymph rig consisting of a small stonefly larva imitation and a trailing fly that imitates a caddis or mayfly. Almost any combination can be effective. I know of a number of anglers who use San Juan Worms

Table 5.1: Water Levels, Targeted Water, Strategies, and Tactics

Water Levels	Targeted Water
Higher than average	Deeper portions of the water column where currents are slower than at the surface. Slower water along current margins, riffle tail-outs, and seam tail-outs. Upstream pockets of holding water behind submerged structure.
Lower than average	Target same holding-water types as that targeted under normal conditions: riffles, seams, confluence points, eddies, flats, and banks with sufficient depth.
Rising water levels	Target same holding-water types as that targeted under normal conditions: riffles, seams, confluence points, and eddies. Pay particular attention to banks and flats.
Receding water levels	Holding water that is threatened with going dry: Target the mouth of side channels, the downstream edge of shallow riffles, drop-off edges of flats, and bankside troughs.

Strategies and Tactics

Larger nymph and streamer patterns that are more visible to trout. Darker patterns are more effective in off-color water, but bright patterns can be effective as well, especially in water that is high but not off-color. Movement, either with materials used in the pattern, with manipulation by the angler (through retrieval, swinging, or jigging), or both.

Slow, measured approach with every kind of pattern.

Use nymphs primarily, especially crustacean patterns on tailwaters.

Nymphs that imitate retreating aquatic invertebrates: stonefly, mayfly, caddis, and chironomid larva, as well as crustaceans and leeches. Use streamers with a wide variety of retrieval speeds and action.

when rivers recede. Worms and leeches have to move to more hospitable parts of a stream just as much as aquatic insects do.

As good as the fishing can be with nymphs, I more often turn to streamers when water levels are dropping. Part of this has to do with the fact that I simply love to fish streamers. Another reason is that fishing streamers in such conditions works. Baitfish will feed on aquatic invertebrates just as much as mature trout and whitefish. As insect larva and crustaceans retreat with receding water, feeding baitfish are available to larger, fish-eating trout. Another reason to use fish streamers when rivers are receding is fish displacement. Baitfish and fingerling trout retreat from their previous lies to find more viable holding water. Much of the time, the holding water that is available is being used by dominant fish. This invasion can cause aggravation among trout already in a lie. The results are vicious attacks on the smaller fish. This is why I use streamers when water levels recede. I target the head of riffles, the edges of flats, and shallow banks. Good things typically happen when I employ this strategy.

CHAPTER SIX

WIND
Using It to Your Advantage

No matter the type of fishing you do, be it for trout, in the salt, for warm-water species, or sea-run fish, wind is seen as the enemy. It is misery to most fly fishers. Accuracy and power disappear, and massive tangles occur on tandem or triple rigs. If you are throwing a streamer, wear Kevlar and be prepared to duck.

Wind is a constant companion for fly fishers. Argentina's Tierra del Fuego is well known for thirty-mile-per-hour sustained winds and gusts up to sixty miles per hour.

Everywhere I have fished, wind has been a factor to one degree or another. When I guided in Tierra del Fuego, Argentina, sustained winds of twenty-five to thirty miles per hour were the norm. Gusts of sixty miles per hour were not out of the question. My excursions to Mongolia, Kazakhstan, Patagonia, Costa Rica, the Bahamas, Belize, and throughout North America have all included wind. It makes those very few days when wind is absent feel that much more special—almost eerily so.

The wind is always there, and we as fly fishers need to accept this fact and not let it ruin our experiences. The first step in accomplishing this is to become a better caster. Some are born with innate athletic abilities. Casting comes easy to these people. But for the rest of us, becoming a better caster means practice, practice, and more practice. Get on the water as much as you can. Have fun and recognize that you are making yourself better at the game.

The second step to dealing with wind is to recognize how it impacts the water you are fishing and to then use the wind to your advantage. That is what this chapter is about.

WIND: ITS CAUSES AND ITS IMPACTS

Wind is caused by a number of meteorological elements. One of these is the jet stream. The jet stream is that portion of the atmosphere where high-elevation winds (above 30,000 feet) are the fastest. Winds near the earth's surface are significantly slower than they are at the heights where the jet stream is, but they are still pretty fast, and they are faster than winds at the flanks of the jet stream. The jet stream is also that line that divides cold air temperatures from warmer ones. In spring, the jet stream is migrating north as our planet tilts. In autumn, it migrates south for the same reason. This is a big reason why many of us in the northern hemisphere experience strong, sustained winds in spring and fall. The movement of the jet stream is fluid and can invoke wild swings in winds on a daily, and sometimes hourly, basis. At times the jet stream can point due north or south before making a turn to either the east or west. At other times it can split, with one arm heading in one direction and the other in another direction before converging hundreds or thousands of miles

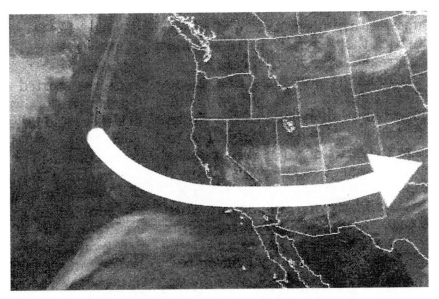

This image shows the jet stream as it flows out of the north Pacific and across the southern portion of the United States. The position of the jet stream suggests strong winds in the Four Corners region and milder winds in the northern Rockies. It also suggests a divergence in temperatures, with cooler air to the north and warmer air to the south. (Image courtesy National Oceanic and Atmospheric Administration)

away. What is important is that the jet stream can tell you where air temperature divergence is occurring and indicate where high winds are likely. The farther the jet stream is from your location, the less likely you will experience high winds.

High winds are also associated with barometric pressure. Winds flow from high pressure to low pressure. Our atmosphere is balancing itself. The bigger the difference in pressure between two systems the faster the winds will be. Thus, you can track the location of high- and low-pressure systems near the waters you are going to fish and have an idea of the wind speeds and its direction.

As I illustrated in chapter 1, changes in barometric pressure play their role in fishing. I experience some of my most-productive fishing during transitional periods, especially when high is shifting to low. Wind and wind speed can assist you in determining when this is occurring. Constant directional shifts in wind and wind speeds indicate a chaotic and

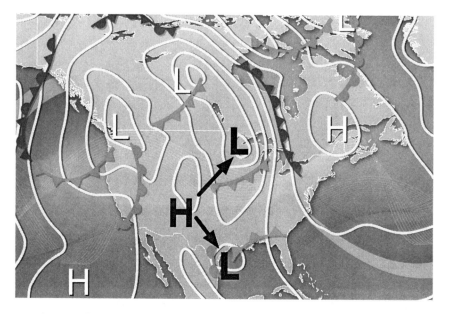

Wind moves from high-pressure systems to low-pressure system as our atmosphere constantly attempts to balance itself. The bigger the difference in pressure systems, the faster the winds will be. You can track the location of high- and low-pressure systems near the waters you fish and get an idea of wind speeds and direction. (© Creativedoc | Dreamstime.com)

inconsistent change in pressure, at least in the short term. When this occurs, fishing becomes less productive. I go to reactionary patterns—big nymphs, big dries moved with authority, and big streamers retrieved aggressively. I will also fish deeper in the water column where trout are attempting to acclimate to the changes.

After moon phase, I find wind to be the least telling of all the natural factors when it comes to finding fish and getting them to take. When I know it is going to be windy, I focus more on what I stated at the beginning of this chapter: adapting to wind and using it to your advantage.

USING WIND TO YOUR ADVANTAGE

I fish and guide on a fair amount of what I call spooky water. This is water where trout are on the defensive due to the characteristics of their environment. Typically the two defining features of such water are shallowness and low gradients that result in calm currents and an almost mirror-like surface. Sometimes there is almost no current at all. Examples of this include storied streams like Silver Creek and the Harriman reach of the Henry's Fork in Idaho and Flat Creek in Wyoming. Lakes I fish that fit this definition are Henry's Lake in Idaho and Hebgen Lake in Montana.

The spookiness of these waters is part of the charm. You can see the fish you are targeting most of the time. They can easily be trophy trout. You might be working toward one for an hour or more. I love this kind of fishing. There is nothing like seeing the behemoths you are casting to.

When the wind kicks up, the surface becomes disturbed. It is more difficult to see fish. But this can be an advantage to the fly fisher. Visibility above the surface becomes difficult for trout in these situations. Apprehension on their part typically subsides. I have experienced very good action in slow, shallow water when the wind disturbs the surface. It has come in handy when I have guided less-experienced anglers who do not have ideal casting skills. The West Thumb of Yellowstone Lake in Yellowstone National Park is a prime example. It has expansive shallow flats along its shores and its islands. There are times when sixty-foot casts to cruising cutthroats are required. Those who lack the ability to get their cast out there, and do so with a soft presentation, can come up empty-handed. However, when the wind picks up and the surface becomes disturbed, I can position my boat closer to the cruising lane. This allows for short casts. Soft presentations are not required because of the unsettled surface. Hookups are much more in the cards.

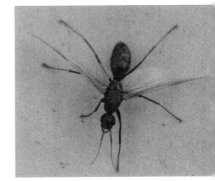

Wind gusts can send grass-hoppers, beetles, and carpenter ants flying onto the surface of streams and lakes, where they become easy pickings for trout. This is one of the lesser-recognized benefits of wind for the fly fisher.

I have experienced these situations on saltwater flats, fishing for sea-run species, and on technical creeks. The wind can help in these shallow, placid environments.

Wind can also help with trout forage. Gusts blow terrestrials like grasshoppers, beetles, and flying ants onto a stream or lake surface, making them available to trout. Aquatic insects with tall wings can be blown over while on the surface, making them easy pickings. Some fly tiers even tie "blown-over duns" for these situations.

Disturbed stream and lake surfaces and insects blowing onto the surfaces are situations caused by the wind that can put the odds in your favor. But nothing beats adjusting your cast.

The easiest way to cast in the wind is to put the wind at your back. This is an inherent skill saltwater guides perform when their guests are having difficult casting into a limiting breeze. Trout guides and anglers do the same.

Most of the trips I guide are done on a river and by boat. They make up a full 70 percent of the days I am on the water. Most of the casts on these trips are made in a downstream direction while drifting with the current. When winds are coming from downstream, they can easily destroy the casts of my guests. Frustration and exhaustion soon follow. When this occurs, we simply stop fishing on the drift. We start to target water where the boat can be stopped or anchored and casts can be made upstream with the wind at our back. Many times we will exit the boat and wade fish upstream on a side channel or long riffle. The same tactic can be used when winds are coming from upstream. I cast to holding water that is downstream of my position using draw-back and parachute casts, described in earlier chapters. These tricks of the trade can save a potentially subpar day.

Putting the wind at your back is an easy remedy. But it isn't always possible. Obstructions may exist behind you that limit your back cast or the water you are targeting may be out of reach. So learning to cast into the wind, or with winds approaching from the left or right, is essential.

One of the first pieces of advice I give to fly fishers who struggle with the wind is to concentrate on the position of the thumb on the casting hand. The standard, and correct, position is for the thumb to be directly behind the handle. This helps create rod balance, accuracy, and power. We are all taught this when we start fly fishing. Nevertheless, I see many anglers lazily slip their thumb off to either the right or left side of the rod.

The standard position of the thumb on the casting hand is on the back of the cork and the opposite side of the reel. This creates balance and power during the cast. When casting into the wind, a hard push of the thumb against the cork during the final phase of the forward cast allows the rod to punch through the wind with a little added power.

The position of the thumb is critical because it can help you "push" the fly through the wind. This is something I witnessed at a casting demonstration given by Joan Wulff several years back. On the forward cast, the thumb on the back of the handle is pushed forward near the end of the power stroke. Wulff describes this as the same kind of motion as pushing in on the button of a screen-door handle. This motion creates a small but noticeable inflection with the wrist, which in turn creates the needed power at the end of the stroke that punches through wind effectively. Wulff illustrates this technique in a number of her instructional videos.

Another tactic that can help fly fishers deal with wind more effectively is to move away from the standard overhead cast and use an angled or side cast. A side cast lowers the rod to an angle that is more in plane with the surface. This action lowers the trajectory of the line. Instead of the line traveling overhead during the casting motion, it is traveling level with the surface. Wind speeds are higher the farther away they are from the surface. A fly line being cast at an angle that is close to the surface is

Whether you are on a stream or a lake, winds have the potential of ruining your day. Fly fishers can use the wind to their advantage by putting it at their back when casting. Another remedy is to keep the rod low and to the side during the cast. Wind speeds are slower the closer they are to the surface. A fly line that is close to the surface is less impacted by wind than a line that is cast overhead.

traveling a path that is less impacted by wind than a fly line that is cast overhead.

Side wind, coming from either the left or the right, is another issue fly fishers face. These winds do not necessarily hinder power and casting distance, but they can impact accuracy by blowing your fly off target. Even worse, these winds can blow your rigging back into you. When side winds occur, and side casts are not possible because of obstacles, you can use a rod tilt on the overhead cast. When the wind is blowing from the right, adjust the rod tip to the right just before the forward cast starts. Perform the same tactic to the left when the wind is blowing to the left. In order to hit your target accurately, you will need to compensate for the wind with rod tilts. These techniques are illustrated in Lefty Kreh's *Casting with Lefty Kreh.*

YOUR GEAR MATTERS: LEADERS, LINES, AND RODS

The type and length of leader you use can help you cast successfully in windy conditions. Manipulating your leader is quick and inexpensive. Traditional tapered-leader material is supple. Its elasticity helps absorb the shock of fish when they run, jump, or throw violent head shakes during the fight. The pliable nature of tapered leader has its limitations, though. Being supple, it has less ability to power through winds during the cast. Tangles and inadequate distance can result. To compensate, many anglers will turn to stiffer, nontapered tippet, like Maxima or Rio's alloy saltwater tippet. A level piece of Maxima or Rio alloy is ideal for streamers. They can also be used for nymphs and dry flies. When the wind becomes too difficult for my guests to cast traditional tapered leader, I will use Maxima to connect their line to their nymph rigs and dry-dropper rigs. This allows their rigging to turn over better in the wind. It also has enough shock absorption to handle the fight after the hookup.

Stiff leader material can help turn your rigging over when casting into strong winds. I will sometimes shorten my leader up, going from my typical length of nine feet down to six or seven feet.

Table 6.1: Wind Direction, Barometric Conditions, and Fly-Fishing Tactics

Wind Direction	Barometric Conditions
Variable	Indicates small shifts and changes in barometric pressure
Wind coming upstream	Can indicate low pressure or barometric pressure in transition
Wind coming downstream	Can indicate low pressure or barometric pressure in transition
Wind coming from the left	Can indicate low pressure or barometric pressure in transition
Wind coming from the right	Can indicate low pressure or barometric pressure in transition

Streams and lakes: More-difficult fishing. Use reactionary patterns and tactics: big nymphs, dry flies, and streamers. Go with nonimitative color. Target deeper portions of the water column where trout are acclimating to pressure shifts.

Streams: Target upstream holding water to put the wind at your back. Target deeper portions of the water column if low pressure dictates. Use disturbed surface to your advantage on low-gradient streams.

Lakes: Position yourself with the wind at your back. Target deeper portions with imitative patterns if low pressure dictates. Use disturbed surface to your advantage.

Streams: Target downstream holding water to put the wind at your back. Target deeper portions of the water column if low pressure dictates. Use disturbed surface to your advantage on low-gradient streams.

Lakes: Position yourself with the wind at your back. Target deeper portions with imitative patterns if low pressure dictates. Use disturbed surface to your advantage.

Streams: Use a rod-tilt cast as described in this chapter. Target deeper portions of the water column if low pressure dictates. Use disturbed surface to your advantage on low-gradient streams.

Lakes: Use a rod-tilt cast as described in chapter 6. Target deeper portions with imitative patterns if low pressure dictates. Use disturbed surface to your advantage.

Streams: Use a rod-tilt cast as described in this chapter. Target deeper portions of the water column if low pressure dictates. Use disturbed surface to your advantage on low-gradient streams.

Lakes: Use a rod-tilt cast as described in chapter 6. Target deeper portions with imitative patterns if low pressure dictates. Use disturbed surface to your advantage.

Another solution for casting in windy conditions is to shorten the leader by a couple of feet. This cannot always be done with nymph rigs if the targeted depth in the water column requires a long leader. However, a short leader can be the answer when fishing surface patterns or streamers. Short leaders will simply turn over easier than longer ones no matter the conditions. Two to three feet is typically all it takes. I have experienced those times when shortening a nine-foot leader to five total feet is needed. If I am dealing with leader-shy fish, I will cut the leader at its bulk end. This will leave the finer end intact where the rigging is attached. Some complain that such a system leaves the flies too close to the line and that this can spook wary trout. But if winds are so strong it requires the manipulation of the leader, then surface currents are likely significantly disturbed. The distance from line to fly is of negligible importance in such cases.

New technologies in rod and line design, not to mention the material used to create them, makes casting in wind a whole lot easier. Most anglers who are new to fly fishing put too much emphasis on rods. Lines are ignored to a certain degree. This attitude has changed quite a bit over the past fifteen years. Much of this has to do with better marketing and better education on the part of instructors, shops, and guides. A good rod-and-line combo assists fly fishers in pushing line through most winds. My favorites include Scott's Radian, Sage's One, and Orvis's Helios matched with Rio's Perception line or Scientific Angler's Sharkwave. Your local fly shop will have all the info on these rods and lines. Local fly shops are great sources of knowledge about gear and the right combination for you. Visit one near you and see what they have to say about their selection. Test their rods and lines and see which combination works best for you. And then make your purchase through them as opposed to an online store. Doing this will help keep fly fishing alive in your community.

CHAPTER SEVEN

READING THE GAUGES
pH Levels, Dissolved Oxygen, and Specific Conductance

On the last day of August 2005, I walked into one of our fly shops located in the village of Moose, Wyoming, after a fun, short, guided trip on the Snake River. The month had been a typically warm one, but our late-summer storms where finally moving in and this particular day featured a midafternoon thunderstorm. I came off the water wet, but happy. The fishing had been good with steady hatches just before the cloud cover rolled in and lasting until we finished up at the boat ramp.

I had been on the water that day with fellow guide Jason Sutton. Sutton is an inquisitive angler, and he was one of the best guides I have ever known. He finished his trip a little before I did. When I walked into the shop, I noticed him sitting at the office computer studying a US Geological Survey (USGS) river gauge website. Sutton waved me over and asked me to look at a particular spike he pointed to on a graph displayed on the screen.

"That rise on the gauge happened right around 2:00," he noted. "Did you happen to notice an uptick in surface feeding at that time?"

I often examined river-flow information on websites maintained by the USGS and the Bureau of Reclamation. But what Sutton was looking at was different. The graph didn't show information on flows at the gauge. Rather, he was looking at a graph displaying pH levels. It displayed constant crests and troughs moving in a consistent pattern across the screen. Then, near the end of the graph, it made a noticeable jump, leveled off, and then came back down to the consistent level it had been earlier.

"That spike is interesting," he said while leaning back in his chair. "Just after that is when the surface action really started to happen for us. And it lasted about as long as it took that spike to come back down and level off."

The gauge that we were studying was one I had looked at dozens of times each year since it came online in the early 2000s. I usually looked at what most fly fishers were concerned with: stream flows (represented by a graph displaying discharge in cubic feet per second) and stream temperatures. These two real-time graphs were at the top of the webpage. Had I bothered to scroll down just a bit more, I would have noticed three other graphs that displayed pH levels, dissolved-oxygen levels, and specific conductance of the water passing by the point of measurement.

Since discovering those graphs more than a decade ago, I have taken the time to study these important factors of streams and lakes

Stream gauges maintained by the US Geological Survey can be found on rivers and creeks throughout the country. Almost all of them provide real-time data on stream flow, height, and temperature. Some also provide information on stream pH, dissolved oxygen levels, and specific conductance. This information can be accessed online via the USGS water page.

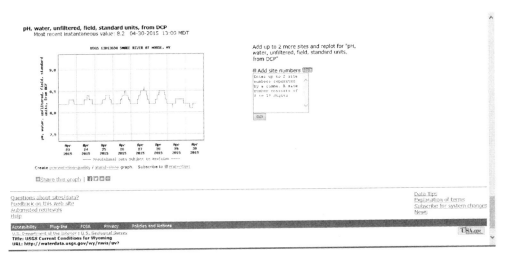

A few stream gauges in the West are equipped with pH, dissolved-oxygen, and specific conductance meters. This graph from the Moose gauge on Wyoming's Snake River shows pH levels during the last week of April 2015. Readings were consistently between 8.0 and 8.5, almost ideal for a healthy trout stream.

where I fish and what they mean to trout and their environment. It has been eye-opening to say the least. By no means are pH, dissolved oxygen, and specific conductance as critical as water temperature, barometric pressure, precipitation, or other elements of fly fishing. But they do play a role, and learning more about them can make you a better angler.

LEVELS OF PH

We all learned about pH in junior high. It is an expression of the acid-base (alkaline) relationship of a particular solution measured on a scale ranging from 0.0 to 14.0. A measurement of 7 represents a neutral solution. A value decreasing below 7 indicates increasing acidity. A value increasing above 7 indicates an increasingly basic solution. An example of a liquid with high acidity would be the juice from citrus fruits, like lemons, limes, and oranges. A liquid with a high basic or alkaline level would be common cleaning solutions like ammonia and bleach.

The pH level influences much in a trout's world. It plays a significant role in egg development, the size and growth of fish as they age, and if they live or die. Numerous studies by fishery biologists have shown that waters high in acidity (pH 5.0 or lower) can be detrimental to fish development and growth, and, in some cases, increases fish mortality. Most of the scientific literature suggests that water leaning toward the alkaline side of the pH scale (from 7.1 to 9.0) is the best condition for trout to obtain large sizes. High alkalinity (pH 10.0 or higher) may not be as detrimental, but it does lead to more lethargy and less feeding activity by fish. Our experiences as fly fishers confirm this for the most part. Lakes and streams with water in the 7.0 to 10.0 range are nutrient rich. They typically have plentiful vegetation, which indicates nutrient rich water. These are also the waters that have a high concentration of the aquatic invertebrates that trout feed on.

Streams and lakes with a pH in the 7.0 to 9.5 range are considered nutrient rich with plenty of forage for trout. This Callibaetis *spinner came from Lewis Lake in Yellowstone National Park, which also contains lots of gray drakes, chironomids, and large numbers of brown trout.*

Waters that dip just below pH 7.0 are typically not a serious problem for trout they hold. In fact, fishing can be very good on these lakes and streams. They can hold a high concentration of trout (although not necessarily a lot of large specimens). It is when they begin to approach 5.0 and lower that conditions get tough. These waters generally have depleted oxygen levels. There is still life on these lakes and streams, but trout populations are not necessarily large, and neither are the individual trout that live in them.

Tolerance levels vary by trout species. Cutthroats, rainbows, and brown trout tend to favor the alkalinity levels between 7.0 and 9.0 and react negatively to levels below 6.0. Some species of char, on the other hand, have a higher tolerance to acidic water. There are strains of brook trout that do just fine when pH is below 5.8. These would be tough conditions for rainbows and cutthroat.

The ecology of a drainage—from its headwaters to its point of termination—plays a crucial role in its acidity and alkalinity. They can vary greatly as its waters move downstream. At the headwaters, acidity is generally high with pH levels at or below 6.0. This is the result of rain and snow that have not captured surface effluents. This is why you won't find large numbers of trout in headwater creeks. It is also the reason why brook trout, with their higher tolerance to low pH, are considered one of the ultimate headwater fish. Some of my favorite backcountry fishing is done on creeks above 8,000 feet where eight-inch brookies are plentiful.

As water continues flowing downstream, it picks up more nutrients. It flows over and through limestone, chalk, basalt, rhyolite, and other carbonate-rich rock. These substances increase the alkalinity and specific conductivity (which is discussed later in this chapter) of the stream. This assists in aquatic invertebrate life, which in turn leads to more and larger fish. We see this clearly on spring creeks. Subsurface spring water not only has regulated temperatures, it also picks up tons of nutrients as it passes through the earth's rock layers. These nutrients come to the surface with cool water. The water temperatures and the pH levels are perfect for trout.

The farther downstream the water goes, the more nutrients it gains. But streams at these lower elevations can have pH levels that exceed 9.5. Nitrate and phosphorus levels, coming from land used for agricultural, industrial, and residential purposes, can be very high. The results are detrimental conditions for trout.

The research on this topic is interesting. It reveals what the big picture is regarding healthy trout water. The more we know about optimal pH levels, and the pH of the lake or river you are fishing, the more we

Streams that flow through carbonate-rich rock have optimum alkalinity. These rivers are high in nutrients and have abundant populations of trout forage like these caddis from the Madison River.

Table 7.1: pH Level, Conditions of Water, and Productivity

pH Level	Conditions	Productivity
0.0 to 5.0	High Acidity	Low productivity. Water is low in oxygen. Limited nutrients with smaller trout. Extremely low levels are not tolerable for trout to survive.
5.0 to 6.9	Low Acidity	High-quality water with good productivity. Lakes and stream can have large concentrations of trout (although not necessarily large specimens).
7.0	Neutral	Indicates pure water. Lakes and streams can have large concentrations of trout.
7.1 to 9.5	Low Alkalinity	High-quality water with good productivity. Lakes and streams can have large concentrations of trout with large specimens.
9.5 to 14.0	High Alkalinity	Low productivity. Not as detrimental to trout as high acidity but leads to lethargic activity and limited feeding. Extremely high levels can be lethal to trout.

know when concerns are raised to protect that piece of water. But this is big-picture stuff. What matters to the fly fisher when he or she is on the water and in the moment is how pH influences trout activity on a given day.

Monitoring immediate pH levels on streams and lakes is difficult. Portable pH meters have been around for quite a while. Some are said to be designed for fishing. I personally haven't had much luck with any of them. The dozen or so times that I have used one yielded either incorrect information or differences from one location to another that seemed inconsequential. Another tool is pH solutions that can be added to a sample of water collected in a vile or petri dish. You match the color of the sample to a color-coded chart, which tells you the pH. I have found this tool to be more accurate. However, taking samples can be time-consuming, and you will have to lug around another two items in your gear bag or pack.

Trial and error has taught me to rely exclusively on stream gauges to monitor changes in pH. What I feel are most important, and what I pay attention to, are pH spikes—rapid upsurges in alkalinity. These spikes can be influenced by a number of factors. Runoff from precipitation is the most common cause (in fact, some die-hard, sea-run anglers in Europe claim it is the surge in pH, and not the increase in water levels, that triggers upstream runs by salmon when they are in rivers to spawn). Fluctuations in water temperature and natural light are also sources of pH spikes.

It is during these times that trout feeding activity intensifies. I also experience increased activity by aquatic invertebrates. This is revealed by either emergences of aquatic insects or by the migration and movement of aquatic insects, leeches, and crustaceans. Instead of using pH meters and sample solutions, I note the time and intensity of the activity. When I finish at the end of the day, I examine the gauges of the stream I am on. It doesn't always occur, but much of the time there will be a noticeable surge in pH, usually between .5 and 1.0 on the graph. Unfortunately, most rivers do not have monitors equipped to measure pH, and I have never been on a body of stillwater that had such a gauge. But for those streams that do, a pH gauge can be a useful device. It can tell you about that body of water, how pH levels impact trout there, and the organisms the trout feed on.

DISSOLVED OXYGEN

As critical as pH level is to the survival, concentration, and size of trout, dissolved oxygen takes precedent in almost every way. Like all water-borne animals, trout will only make it if there is sufficient oxygen in their environment. Fluctuations dictate how active they will be. If those levels reach certain thresholds (both on the high end and the low end), trout can perish.

Dissolved oxygen refers to the amount of oxygen that is being carried in a measurement of water. Levels are measured in milligrams per liter (mg/L). Most scientists who study dissolved oxygen—people like hydrologists, geologists, and biologists—measure dissolved oxygen on a scale with a range from 1.0 mg/L to a high of 14.5 mg/L. Trout and trout forage like stoneflies and mayflies require a minimum of between 5.0 and 6.0 mg/L of dissolved oxygen to survive. Some aquatic invertebrates like chironomids, some caddis, and leeches can survive in water with far less than 6.0 mg/L of dissolved oxygen. The higher the amount of oxygen saturation (another term for dissolved oxygen) in water, the more beneficial conditions are for trout. Dissolved oxygen levels of 9.0 mg/L to 12.0 mg/L can generally support a healthy, abundant, sustainable population of trout.

A number of factors contribute to the level of dissolved oxygen in water. Aeration caused by the churning and running of currents over a streambed is a prime contributor. We see this on riffles and among submerged structure. It is one of the reasons, along with protection and proximity to forage, that trout favor these types of holding water. On stillwater, this kind of aeration is not possible. But lakes can generate ample dissolved oxygen through photosynthesis, the upwelling of subsurface springs, and the churning of surface water by winds.

Water temperatures assist in maintaining levels of dissolved oxygen. The cooler the water, the better oxygen is retained. When water is at 71 degrees, it contains 9.0 mg/L of dissolved oxygen. When water is at 50 degrees, it contains 11.5 mg/L of dissolved oxygen. At 46 degrees,

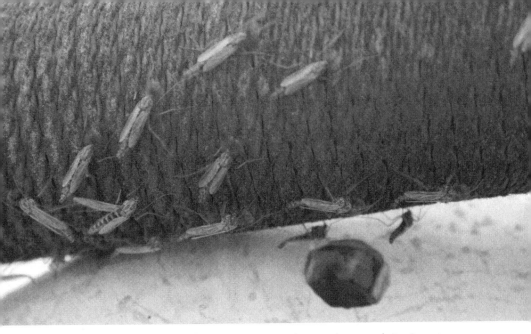

Chironomids are one of the few aquatic invertebrates that can thrive in water with less than 6.0 mg/L of dissolved oxygen.

12 mg/L of dissolved oxygen (the upper level of optimum oxygen levels) is attained.

Water levels can also play a role. Streams and lakes with low water levels can experience considerable fluctuations in water temperatures. These conditions can lead to rampant vegetation growth. Because of the role aquatic vegetation can play in controlling the level of oxygen and other compounds, massive swings in dissolved oxygen can result. These changes can occur from one part of the day to another.

The consumption of oxygen by trout varies with stream temperatures. When water temperatures are in the low to mid-40s, trout use between 50 and 80 mg of oxygen per hour. But as temperatures approach the high 60s to low 70s, they need 200 to 300 mg every hour. When you consider that trout require large amounts of oxygen at these temperatures, their metabolic rates are rapid when water temperatures are high, and there is less oxygen in the water to begin with, it is easy to understand the extreme stress trout can be under. State and federal agencies can shutdown fishing in trout streams in the western US when water temperatures surpass 70 degrees.

Examining dissolved oxygen on trout streams and lakes is as cumbersome and time-consuming as checking pH levels. There are several

The Firehole River in Yellowstone National Park is fed partially by thermal features along its entire length. It can fish great all year long, but some years, temperatures can reach well into the 70s. Fishing can be shut down entirely by the National Park Service, primarily because the fight after being hooked is lethal to trout.

types of oxygen meters on the market, but they can be quite expensive—more than $400 in most cases—and they are designed more for controlled environments like hatcheries and research labs then they are for anglers in the field. Many of the stream gauges that have pH monitors also have devices that measure dissolved oxygen levels. These are what I look at. They exist on only two of the rivers I fish regularly in Idaho and Wyoming. It would be nice to see them on more waters throughout the country. On those waters that don't have dissolved oxygen meters, I will keep a record of water temperatures throughout a day of fishing. These two variables mirror each other closely.

Knowing dissolved oxygen levels can tell you when and if trout are feeding, as well as where they might be holding. There is a close relationship between air temperatures, water temperatures, and oxygen. When I examine dissolved oxygen levels, they are usually higher in the morning and slowly drop as the day progresses. Its line on the graph follows the inverse progression of water and air temperature through the day: oxygen is increasing as air and water temperatures drop.

Feeding activity is best in the morning and slows as oxygen levels drop below 8.8 mg/L later during the day. I target the extreme head of riffles and confluence points when this occurs. Aeration and oxygen generation is greatest at the head of these holding-water types. I also target the deeper portions of riffles, pools, eddies, and seams with subsurface patterns. These waters allow for greater depth in the water column. Cooler water exists at the deeper parts of the water column. This is where there is better oxygen retention.

While stillwater bodies rarely have oxygen monitors, fly fishers can keep track of levels by measuring water temperatures. As water temperatures rise, dissolved oxygen levels drop. Temperatures will often be warmer at the surface and the top few feet of the water column unless it is immediately after spring ice-out. When these temperatures hit 68 degrees

This rainbow was landed on the Madison River in Montana on a cold February day. Water temperatures were hovering around 41 degrees. Feeding by trout is minimized when water temperatures are cold, and they have limited reserves. However, cold water retains oxygen, and chances for recovery are maximized. The fly fisher can help by getting trout in fast and releasing them immediately.

A wide range of specific conductance can produce good environments for game fish and good fishing for the angler.

or more, I begin to fish deeper in the water column, going below the four-foot level and sometimes below the twelve-foot level; higher levels in the water column are devoid of feeding fish. Oxygen supplies are greater at these depths than they are closer to the surface, and trout will congregate there. It is striking that fishing can be excellent on lakes when temperatures at the top of the water column are from 64 to 66 degrees. Yet when temperatures are just a couple of degrees warmer, action can shut down completely.

SPECIFIC CONDUCTANCE

Specific conductance refers to the ability of water to conduct (move) an electric current between two points. Water that is high in dissolved solids moves current better and faster than water that is low in dissolved solids. Specific conductance is measured in microsiemens per centimeter (mS/cm).

Table 7.2: Dissolved Oxygen Levels, Trout Activity, and Fly-Fishing Strategies and Tactics

Dissolved Oxygen Levels	Trout Activity	Strategies/Tactics
9.0–12.0	Optimum levels for trout. Feeding and holding can occur in all possible holding-water types and parts of the water column. Extended exposure to dissolved oxygen levels at or above 11.0 could be harmful to trout.	**Streams:** Target all holding-water types where trout can conceivably hold and feed, with special attention paid to the surface and top of the water column. **Lakes:** Target all holding-water types where trout can conceivably hold and feed, with special attention paid to the surface and top of the water column.
7.5–8.9	Stress increases for trout due to limited oxygen and higher water temperatures. Trout begin to hold in areas where oxygen generation is greatest and water temperatures are coolest.	**Streams:** Target the extreme head of riffles and confluence points with subsurface patterns. Fish subsurface patterns in deep parts of the water column of riffle pools and the tail of seams. **Lakes:** Fish deeper portions of the water column (below the four-foot level) at drop-offs and ledges with subsurface patterns.
Below 7.5	Extreme stress for trout. Trout and other fish will hold at the bottom of the water column or at those portions where cool water is entering a stream or lake.	Stop fishing until dissolved oxygen increases to levels safe for trout.

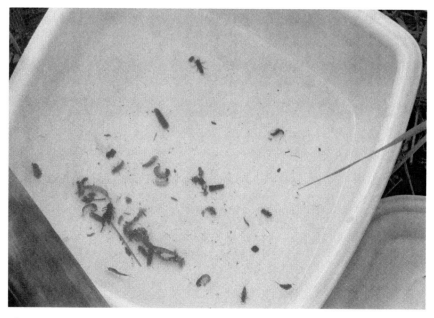

The Henry's Fork has ideal specific conductance and pH levels, as is evident from its high concentration of aquatic invertebrates.

Why is this important? Can it really mean that much to the fly fisher? In comparison to the other factors, it is not critical. But it is worth understanding specific conductance. It is a good indicator of water quality and quality fishing conditions.

Waters that have extremely high conductivity are deemed to have high levels of pollutants and impurities and, thus, are not healthy for fish. Conversely, streams that have extremely low conductivity have very low levels of nutrients. Streams and lakes in this category are not ideal for large numbers and sizes of trout.

A number of natural and unnatural components influence specific conductivity in water, and ranges of these components can vary greatly from one piece of water to another. For example, some estuaries in coastal fisheries exhibit conductance at several thousand mS/cm. These waters are impacted by salinity from ocean currents flowing into them during high tides. This factor contributes to high conductance. Despite this, these bodies of water can fish exceptionally well for everything from saltwater sport fish to sea-run salmon. Some even fish well for resident trout.

In contrast, the upper Truckee River on the eastern slope of the Sierra Nevada Range exhibits specific conductance between 30 and 150 mS/cm. The Truckee has a solid reputation as a trout stream.

Rivers like the Truckee can experience dramatic swings in specific conductance throughout the year. Spring runoff and the ramping up of releases from dams spill high amounts of natural effluents into streams. This substantially raises levels of specific conductance. As runoff subsides or dam releases are ramped down, specific conductance levels come down just as fast.

At the same time, some trout streams and lakes have stable levels of specific conductance due to their stable flows and chemical composition. Examples include the famed chalk streams of the United Kingdom, the limestone streams of the eastern United States, and the Henry's Fork in Idaho, which is surrounded by basalt and rhyolite. The beds of these streams are composed of calcium carbonate, magnesium carbonate, and/or dolomite. These minerals leach into the water, substantially increasing its nutrient level. Such minerals constitute the dissolved solids that give water high specific conductivity. These streams are prime cases of water that is both high in specific conductance and supports large numbers of trout.

Specific conductance isn't a factor I keep track of often. I typically examine USGS graphs (once again, those gauges that have pH and dissolved oxygen monitors will also measure specific conductance) on the water I fish when stream levels are fluctuating or there are dramatic shifts in water temperature. My experience is that production doesn't vary much based on changes in conductance. I do, however, believe it is a factor worth keeping track of over the long term. Changes in specific conductance could suggest changes in the quality of the water you fish. Those changes could be beneficial. They could also be signs of something detrimental.

CHAPTER EIGHT

AS IF WE REALLY KNOW
Reading the Lunar Phase

I'm a moon fisherman. And fishing isn't going to be worth a shit tomorrow."

This was the emphatic statement made by a well-worn, seventy-something angler at Headhunter's Fly Shop in Craig, Montana. It was the third week of April, and I was having a short break from early season

The moon rises over a lone fly fisher on a windswept river in Argentina. There are countless stories of the role lunar phase plays in fish behavior. Nonetheless, we don't have a complete understanding of the moon's influence.

guiding in my neck of the woods to take in a couple days of fishing on the Missouri River. I had been talking to a member of the shop staff about conditions for the following day, particularly about the unusually warm temperatures and sunny skies to be expected, when the old-timer decided to chime in.

He went on a quiet but self-assured rampage. The old man gave us all kinds of reasons for why weather, pattern selection, holding water, and time of year didn't matter in the least. Everything, according to this gentleman, revolved around the moon. When it was full, when it was new, when it was rising, and when it was setting, was all that mattered. A fly fisher who is on the water when the lunar phase wasn't right is an idiot. That was his opinion of me for sure: the moon wasn't right, and I was going to be on the water.

The next day started slow, but by lunch my friends and I were hammering fish with nymphs and streamers, and we ended the afternoon by picking up several fish on the surface with blue-winged olive and caddis emergers.

So much for this guy's opinion of fishing when the moon wasn't right.

Lunar phase is one of the greatest mysteries in the world of fly fishing. Its impact on saltwater angling is well documented. The moon's influence on saltwater environments is significant due largely to its effect on rising and falling tides. In almost every other kind of fly fishing, however, the role the moon plays in terms of fish behavior is not backed up by the scientific literature. This goes for everything from warm-water to sea-run to freshwater fly fishing. What we are left with are the stories, observations, and experiences of those who pay attention to it.

Of all the factors I cover in this book, moon phase is probably the one that I pay attention to the least. There was a period of approximately ten years when I kept lunar records, but I found very little consistency in my fishing based on the moon's phase. Today, I only keep mental records. Climatological dynamics and water factors are what matter most. But while its role is not convincing to me, lunar phase deserves mention in a work like this.

MOON PHASE: WHAT WE THINK AND WHAT WE KNOW

Speculation on the role lunar phases have on living creatures is wide and varying. Psychologists note that criminal and psychotic behavior rise during a full moon. There is also evidence of sleep disruption and increased incidents of dog bites requiring hospitalization. Biologists note that some bugs dig deeper holes when the moon is full, and that lions, which typically kill prey at night, will intensify their daytime kills immediately after a full moon. Some even suggest an association between the full moon and fluctuations in the stock market. Nonetheless, those who study the matter have found little, if any, link between moon phase and alterations in behavior.

We fly fishers are certain of an association between lunar phase and fish behavior. There are numerous stories handed down through the generations regarding this topic. One that is widely held in the Yellowstone region is that brown trout leave their lake environments to spawn in their natal streams on the first full moon after autumn equinox. This

The autumn spawning run of brown trout in the Rocky Mountain West is a much heralded event by local and visiting fly fishers. Many anglers in the Yellowstone region contend that these runs begin on the first full moon after equinox.

Many anglers believe that slow fishing during a full moon has less to do with trout being able to see at night and more about the impact of lunar pull on magnetic particles in the cells of fish. This may be the reason many of us experience slower fishing when nighttime moonlight is obscured by cloud cover. Nonetheless, most fly fishers get on the water no matter what part of the lunar cycle we happen to be fishing in.

contradicts the findings of fishery biologists whose studies show that brown trout–migration distances are shortest during full moon periods.

Perhaps the most cited influence, and the one that is easiest to explain, involves feeding: fish feed at night when there is a full moon because they can see forage better. This leads to more lethargic feeding behavior during daylight hours because they are sated.

This is a simple explanation that makes a lot of sense. My observations, however, suggest that there is very little consistency with this purported causal relationship. The actual amount of feeding that occurs at night is a good example. Much of the guiding I have done over my life includes overnight trips in the Yellowstone backcountry and on the South Fork of the Snake River in Idaho. These excursions give me ample opportunity to fish at night, including nights just before, during, and after a full moon. Some of these nights are full of fun surface action. There was also good action on streamers just below the surface. But there were also nights when the bite was subpar. And on those nights when the bite was on, fishing the next day was not always dreadful. Most of the days were fairly good.

Many fly fishers who believe in an association between lunar phase and trout behavior contend that it is not the ability to see at night that is the cause. Rather, it is the pull of the moon on the earth that triggers reactive behavior. Here is where there is at least a little evidence of a causal relationship. When the moon enters a new phase, there is a shift in the earth's electromagnetic field. Salmon have magnetic particles in their lateral line. Some recent research shows that Pacific and Atlantic salmon use these particles to orient themselves to our planet's magnetic field. This ability allows salmon to navigate their way back to their home rivers to spawn. Trout have these same magnetic particles (called magnetite) in their cells. These cells are believed to be used by trout to navigate long distances in streams and lakes.

Could it be this play between magnetic particles in trout and the shifts in earth's magnetic field that explains their behavior during a full moon? Anglers who believe this to be true point out that there is limited activity during a full moon even when there is substantial cloud cover that limits natural light at night. There is little if any scientific evidence backing up this argument. Nevertheless, I believe this argument has more merit than those who claim trout are eating more at night because they can see better.

STOP WORRYING SO MUCH AND FISH!

Regardless of our convictions about lunar phase and its influence on fish behavior, the truth is we have little scientific data to back up an association. Most of us are guilty of confirmation bias, where our selective thinking leads us to seek and find information that confirms our beliefs and ignores evidence to the contrary. But we do have our beliefs regarding this mysterious part of the fly-fishing world, and those beliefs can be hard to let go of. Some contend that it is dawn and dusk that produce best when the moon is full. Others claim that we should just avoid fishing during a full moon altogether. I try not to worry about it. As Denny Rickards has said, "Trout can and are caught during a full moon, but only when it is right for them, not me."

And if the fishing sucks, you can always blame it on the moon.

PUTTING IT ALL TOGETHER

In the introduction to this book, I state how, by its very nature, fly fishing is an inquisitive game. In this way it has much in common with other sports, like skiing and surfing. Some skiers and surfers can become so fascinated in the performance of their tools that the design of their boards, skis, and tuning devices becomes a central focus. Skiers and surf-

A number of natural factors influence production on stillwater, including barometric pressure, precipitation, natural light, and water temperatures. Fly fishers who understand the impact of these factors have a leg up in catching more and larger trout.

ers can also become enamored with the study of snow science, hydrology, and weather patterns. This allows them to learn more about their sport and have fun doing it.

So, too, it is with the fly fisher. Better performance on the water comes from those in our sport who become infatuated with the design of rods, lines, and reels, and find ways to improve their function. The study of hydrology or entomology captivates some anglers, and they use what they learn to have success on the water. This type of intellectual stimulation—be it the design of our tools, studying entomology, or becoming more knowledgeable of stream hydrology—allows us to learn more about this game we play.

I have always been intrigued by the aspect of holding water and how it varies from river to river or lake to lake based on stream gradient, nutrient levels, elevation, trout species, and other factors. When you guide and fish on such varied water throughout the West as I do, it is easy to see why holding water can become the focal point for an angler. This is the intellectually stimulating part of the game that I find to be most enjoyable. It is also among the most important elements of the sport if you want to catch fish.

The factors I examined in this book—barometric pressure, precipitation, natural light, water temperature and levels, wind, pH and dissolved oxygen, and moon phase—are an extension of my fascination with holding water. They play a significant role in terms of where trout hold, how they behave, the forage they are eating, and how they are feeding. Having knowledge about these factors and how they impact trout behavior can give you another tool for greater success on the water. Knowing how trout react to changes in water temperature or stream flows assists you in finding where they are horizontally across types of holding water like riffles, seams, eddies, flats, drop-offs, troughs, banks, and structure. This knowledge also helps you find where trout are holding vertically in the water column.

Consider this book's analysis of precipitation and its impact on trout water. Understanding how precipitation sparks emergences of mayfly species can help you determine the right time to switch from your trusty attractor pattern to a PMD or a blue-winged olive. Natural light

Fishing stillwater can be a perplexing task due to the vast number of factors that must be considered. The influence of water temperature, barometric pressure, natural light, and lunar phase cannot be understated. Having a grasp of these factors and their impact on trout can lead to more success on the water.

conditions is another good example. Recognizing how natural light impacts what trout see can help you make the right pattern selection at the right time. Knowing it all can be overwhelming if not impossible. If you have a grasp of those factors that you feel are important, however, it will lead to better results on the water.

The perplexing piece of the puzzle comes with the interplay occurring between different factors. I typically experience good fishing when the barometer is dropping. A dropping barometer is oftentimes accompanied by precipitation. When it starts to rain, the fishing on the streams I am on the most can be off the charts with dry flies and streamers. But which of those two factors makes the fishing so good so many times? Is one of them the true answer for increased production, or is it the interaction between the two?

Sometimes there are factors that completely butt heads. Imagine fishing an off-color stream under sunny skies. You will remember the sayings "dark water—dark fly" and "bright sky—bright fly" from earlier chapters. But what do you do if you have off-color water on a bright day, or gin clear water on a cloudy day? What color do you go with then? The best that a fly fisher can do is experiment and see if a pattern develops. Again, this is what makes studying external factors so much fun.

And then there are those factors that more or less align with each other. For example, cold water is generally high in dissolved oxygen while warmer water is low in dissolved oxygen. In this case, it is water temperature that dictates oxygen levels. Alternatively, streams that have optimum pH levels also have good specific conductance. In this case, one isn't necessarily influencing the other. Rather, they are both being influenced by factors within the environment, like the element-laden rock streams flows through and runoff from precipitation or rainfall.

In the end, all trout water is different. The best advice I can give fly fishers is to not fish one piece of water the same way they fish another. I certainly don't. I know from experience that trout on a particular stream or lake might react to changes in barometric pressure and pH levels in a way that is completely different from trout on another body of water only a few dozen miles away. Sometimes it is a difference in elevation, other times it is a difference in trout species, and other times it is a difference in stream gradient and nutrient levels.

It is my hope that this book gives you a template for analyzing the waters you fish with a new per-

Knowledge of holding water and trout forage is critically important to the fly fisher. Understanding external factors, however, can take your fishing to another level.

spective. The factors I examined matter, and knowing more about them can matter, too. Having this knowledge can take your fly fishing to the next level. It is a critical step in that constant progression of becoming a better angler.

BIBLIOGRAPHY

Allen, Boots. *Modern Trout Fishing: Advanced Tactics and Strategies for Today's Fly Fisher.* Guilford, CT: The Lyons Press, 2013.

Baldwin, Casey M., David A. Beauchamp, and Chad P. Gubala. "Seasonal and Diel Distribution and Movement of Cutthroat Trout from Ultrasonic Telemetry," *Transactions of the American Fisheries Society* 131, no. 1 (2002): 143–58.

Behnke, Robert J. *Trout and Salmon of North America.* New York: The Free Press, 2002.

Budy, Phaedra, Sara Wood, and Brett Roper. "A Study of the Spawning Ecology and Early Life-History Survival of Bonneville Cutthroat Trout," *North American Journal of Fisheries Management* 32, no. 3 (2012): 436–49.

Bunnell, David B., J. Jeffery Isley, Kyle H. Burrell, and David H. Van Lear. "Diel Movement of Brown Trout in a Southern Appalacian River," *Transactions of the American Fisheries Society* 127, no. 4 (1998): 630–36.

Clark, Michael L., and Louis A. Helfrich. "Comparison of Water Quality and Rainbow Trout Production in Oxygenated and Aerated Raceways *North American Journal of Aquaculture* 68, no. 1 (2006): 41–46.

Coughlin, David J., and Craig W. Hawryshyn. "The Contribution of Ultraviolet and Short-Wavelength Sensitive Cone Mechanisms to Color Vision in Rainbow Trout," *Brain Behavior Evolution 43, no. 4–5 (1994): 219–32.*

Curry, Reed F. *The New Scientific Angling: Trout and Ultraviolet Vision.* n.p.: Buckram Publishing, 2009.

Daniel, George. *Dynamic Nymphing: Tactics, Techniques, and Flies from Around the World.* Mechanicsburg, PA: Stackpole Books, 2012.

Hafele, Rick, and Dave Hughes. *Western Mayfly Hatches: From the Rockies to the Pacific.* Portland, OR: Frank Amato Publications, 2004.

Hansen, Adam G., David A. Beauchamp, and Erik R. Schoen. "Visual Prey Detection Responses of Piscivorous Trout and Salmon: Effects of Light, Turbidity, and Prey Size," *Transactions of the American Fisheries Society*142, no. 3 (2013): 854–67.

Hartman, Kyle J., and Michael A. Porto. "Thermal Performance of Three Rainbow Trout Strains at Above-Optimal Temperatures," *Transactions of the American Fisheries Society* 143, no. 6 (2014): 1445–54.

Kageyama, Colin J. *What Fish See: Understanding Optics and Color Shifts for Designing Lures and Flies.* Portland, OR: Frank Amato Publications, 1999.

Knopp, Malcolm, and Robert Cormier. *Mayflies: An Angler's Study of Trout Water Ephemeroptera.* Helena, MT: Greycliff Publishing, 1997.

Kreh, Lefty. *Casting with Lefty Kreh.* Mechanicsburg, PA: Stackpole Books, 2008.

Matthews, K.R., and N.H. Berg, "Rainbow Trout Responses to Water Temperature and Dissolved Oxygen Stress in Two Southern California Stream pools," *Journal of Fish Biology* 50, no. 1 (1997): 50–67.

McMahon, Thomas A., Beth A. Bear, and Alexander V. Zale. *Comparative Thermal Preferences of Westslope Cutthroat Trout and Rainbow Trout*," Bozeman, MT: Wild Fish Habitat Initiative, 2006.

Monmonier, Mark. *Air Apparent: How Meteorologists Learned to Map, Predict, and Dramatize the Weather*. University of Chicago Press, 1999.

Ostrand, Kenneth G., and Gene R. Wilde. "Changes in Prairie Stream Fish Assemblages Restricted to Isolated Streambed Pools," *Transactions of the American Fisheries Society* 133, no. 6 (2004): 1329–38.

O'Neal, Sarah L., and Jack A. Stanford. "Partial Migration in a Robust Brown Trout Population of a Patagonian River," *Transactions of the American Fisheries Society* 140, no. 3 (2011): 623–35.

Rader, Russell B., Timberley Belish, Michael K. Young, and John Rothlisberger. "The Scotopic Visual Sensitivity of Four Species of Trout: A Comparative Study," *Western North American Naturalist* 67, no. 4 (2007): 524–37.

Randall, Jason. *Feeding Time: A Fly Fisher's Guide to What, Where, and When Trout Eat*. Mechanicsburg, PA: Stackpole Books, 2013.

Rosenbauer, Tom. *The Orvis Guide to Prospecting for Trout: How to Catch Fish When There's No Hatch to Match*, rev. ed. Guilford, CT: Lyons Press, 2000.

Ross, David A. *The Fisherman's Ocean: How Marine Science can Help You Find and Catch More Fish*. Mechanicsburg, PA: Stackpole Books, 2000.

Slavík, Ondřej, Pavel Horký, Tomáš Randák, Pavel Balvín, and Michal Bílý. "Brown Trout Spawning Migration in Fragmented Central European Headwaters: Effect of Isolation by Artificial Obstacles and the Moon Phase," *Transactions of the American Fisheries Society* 141, no. 3 (2012): 673–80.

Thomason, Arlan. *Bug Water: A Fly Fisher's Look through the Seasons at Bugs in Their Aquatic Habitat and the Fish that Eat Them.* Mechanicsburg, PA: Stackpole Books, 2010.

Toregsen, Christian E., Joseph L. Ebersole, and Druscilla M. Keenan. *Primer for Identifying Cold-Water Refuges to Protect and Restore Thermal Diversity in Riverine Landscapes.* Seattle, WA: US Environmental Protection Agency, February 2012.

Utz, Ryan, and Kyle Hartman. "Temporal and Special Variation in the Energy Intake of a Brook Trout (*Salvelinus Fontinalis*) Population in an Appalachian Watershed", *Canadian Journal of Fisheries and Aquatic Sciences*, Vol. 63: 2675-2686, 2006.

Wedemeyer, Gary, ed. *Fish Hatchery Management.* Bethesda, MD: American Fisheries Society, 2002.

Wehrly, Kevin E, Lizhu Wang, and Matthew Mitro. "Field-Based Estimates of Thermal Tolerance Limits for Trout: Incorporating Exposure Time and Thermal Fluctuation," *Transactions of the American Fisheries Society* 141, no. 6 (2006): 1433–38, 2006.

Whitlock, Dave. *Dave Whitlock's Guide to Aquatic Trout Foods,* 2nd ed. New York: Lyons Press, 2007.

Williams, Jack. *The Weather Book: An Easy-to-Understand Guide to the USA's Weather.* New York: P Vintage Books, 1997.

Witschi, William A., and Charles D. Zeibell. "Evaluation of pH Shock on

Hatchery-Reared Rainbow Trout," *The Progressive Fish-Culturist*, Vol. 41, no. 1 (1979): 3–5.

Woodmencey, Jim. "What Makes the Wind Blow?," *Jackson Hole News and Guide*, May 14th, 2014.

INDEX

Page locators in *italics* indicate photographs.

daily fluctuations in water temperature, 55

damselfly larva imitations: and fishing productivity, 78; Fur Damsel, 28; retrieval speeds, 28, 29

Daniel, George, 98

"dark water–dark fly" mantra, 100–101, *101*

Dave Whitlock's Guide to Aquatic Trout Foods (Whitlock), 84

Day-2 Midge Pupa, 67, *67*

deepening troughs and barometric pressure, 19

degreasing agents for nymph rigs, 71

dehydration and mayfly emergence, 49, *49*

descending water patterns, 106–7

dissolved oxygen, 130–34, 135

Dornan, Will, 43, 101

Double Humpy, 87

dragonfly larva imitations, 29

drizzle, 37–38

dry air and barometric pressure, 18

dry flies: falling barometric pressure and fishing productivity, 28; flurries and fishing productivity, 40, 41–42; high barometric pressure and fishing productivity, 23; light rain and fishing productivity, 39; moderate to heavy snowfall and fishing productivity, 43, 45; pattern size and color, 83, 84–85, 87, 89, 92, *92*; and warm water conditions, 61–62. *See also* presentation; retrieval movements and speeds

Eagle Lake, 22

eddy seams, cold-water fishing technique, 66

electric current, specific conductance, 134, 136–37

entomology, 9–10

Erikson, Pete, 71

falling barometric pressure and fishing productivity, 27–29, *27, 28, 37*

Feeding Time (Randall), 57

Firehole River, 53–54, 132, *132*

fish behavior: air bladders and barometric pressure, 20–22, 24; moderate barometric pressure and fishing productivity, 25–26, *26*; regional differences in, 21; rising barometric pressure and fishing productivity, 24–25, *24*

Fish Hatchery Management (Wedemeyer), 52

The Fisherman's Ocean (Ross), 21

fishing strategies and tactics: barometric pressure and weather conditions summary table, 32–33; falling pressure conditions, 27–29, *27, 28*; high

Bug, 105, *105*; moderate to heavy snowfall and fishing productivity, 44; receding water levels, 106–7; snow flurries and fishing productivity, 40–42; tungsten bead nymphs, 70–71, *71*; and warm water conditions, 62; water depth and temperature stratification, 60–61; water temperature and fishing depths, 70–71, *71*

off-color, high-flow conditions, 95, *95*, 96, 97, 99–102, *99, 100, 101*
olive comparaduns, 41–42
Onchorhynchus genus, 52–53
Owyhee River, 102
oxygen, dissolved oxygen, 130–34, *135*

pale morning duns, 78
Pat's Rubber Leg, 99, 101, *101*, 102
pattern size and color: high-flow, off-color conditions, 95, *95*, 97, 99–102, *100, 101*; natural light conditions, 83–87, *84, 85, 86*, 89, 92, *92*
pH and fishing productivity, 48, 123–30, *125, 128*
Pheasant Tail Nymphs, 49
PMD Emerger, *50*
PMDs (pale morning duns), 40, 49–50, *50*

poikilothermy, 56–60, 62, 70–71
pools, warm-water fishing technique, 66–67
precipitation: drizzle, 37–38; and fishing productivity, 35–37, 45, 48–50; fishing strategies and tactics summary table, 46–47; hail, 40; light rain, 38–39; moderate to heavy rain, 39; moderate to heavy snowfall, 43–45, *43*; moist air and barometric pressure, 18; sleet, 39–40; snow flurries, 40–42, *41*; snow showers, 42–43, *42*
presentation: dry flies and high atmospheric pressure fishing productivity, 23; low-water conditions, 102–3, *102*; moderate barometric pressure and fishing productivity, 25–26; windy conditions, 103–4, *104*, 111–12, 115–18, *117, 118*
Purple Haze, 87

Quad Bunny, 87

rain: drizzle, 37–38; light rain, 38–39; moderate to heavy rain, 39
rainbow trout: cold water adaptation, 52–53, *53*; Madison River, 97, *133*; natural light conditions and feeding behaviors, 81–82